LIGNIN BIODEGRADATION AND TRANSFORMATION

LIGNIN BIODEGRADATION AND TRANSFORMATION

RONALD L. CRAWFORD

University of Minnesota
Department of Microbiology and
Gray Freshwater Biological Institute
Navarre, Minnesota

A WILEY-INTERSCIENCE PUBLICATION

JOHN WILEY AND SONS

New York · Chichester · Brisbane · Toronto

Library of Congress Cataloging in Publication Data

Crawford, Ronald L.   1947-
  Lignin biodegradation and transformation.

  "A Wiley-Interscience publication."
  Includes index.
  1. Lignin—Biodegradation. 2. Biotransformation
(Metabolism) I. Title.

QK898.L5C7   581.87'5   80-39557
ISBN 0-471-05743-6

Printed in the United States of America

10 9 8 7 6 5 4 3 2 1

To Onie and Lisa

FOREWORD

Research concerned with various aspects of the microbiological degradation of lignin in plant materials has appeared in the scientific literature for many years. Soil microbiologists have been interested in the rates of turnover of lignin in nature and in the roles of lignin, or its breakdown products, in the process of humification in soil. Microbiologists have also long been interested in identifying the microbial types that play a part in lignin biodegradation. The physiology of lignin catabolism by microorganisms has also intrigued microbiologists, who have attempted to elucidate the pathways by which this extremely complex aromatic polymer is decomposed into smaller fragments and eventually to CO_2 and H_2O. In recent years researchers have also begun to recognize the value of lignin, one of Earth's most abundant renewable resources, as a potential source of useful industrial chemicals. As a result, research into the degradation and transformation of lignin by microorganisms has recently become of major interest to researchers interested in developing alternative, renewable sources of chemical feedstocks.

Still, the microbiology of lignin degradation is not well understood. Until recently, only certain fungi were known unequivocally to degrade lignin. The pathways of lignin degradation had not been even crudely elucidated and little was known about enzymes that attack polymeric lignin. Indeed, much of our knowledge of lignin biodegradation was based on conjecture supported by limited research data. The reasons for this lack of basic information, despite decades of research, are numerous. However, chief among them is the fact that the structure of lignin was not well established until fairly recently. As that structure has been slowly worked out, it has become clear that lignin's biological degradation must indeed require a multiplicity of specific enzymes capable of attacking numerous diverse structural elements. To confound the situation, available assays for quantifying lignin degradation have been insensitive, subject to inaccuracies because of interference by other biological materials, and often prone to misinterpretation by researchers.

Recently, however, major advances in lignin biodegradation research have occurred. A substantial achievement has been that lignin's basic chemical structure, as a complex phenylpropanoid polymer with a variety of intermonomeric linkages, has been firmly established. As a result, methods for the comparative chemical analysis of natural and microbially degraded lignins have been worked out. In another major advance, very sensitive and unequivocable assays for lignin degradation have been developed. In turn, the range of microorganisms known to attack lignin has grown to include a variety of bacteria and fungi not previously regarded as lignin decomposers. Even the pathways of lignin degradation are becoming known, and the elucidation of the enzymology of lignin degradation and transformation is on the horizon. In other words, our knowledge of lignin degradation is growing by leaps and bounds.

The current rapid accumulation of knowledge has created a problem in the sense that many workers have not been able to keep abreast of continuing advances. In particular, researchers are lagging behind in their knowledge and use of currently available methodologies. There is a real need for an up-to-date and detailed summary of current knowledge. Several recent journal reviews on lignin degradation can help lead workers to the appropriate literature, but they do not provide researchers with such things as insights into the often subtle influences of experimental variables on research results. Neither have these reviews adequately pieced together, into a single written work, all the diverse bits of information that have accumulated concerning lignin chemistry and biosynthesis and microbial lignin degradation. Such a complete analysis is needed because it is this kind of detailed study of the subject that can show how many individual research findings are relevant to one another in a broader sense.

As I read this book, I was impressed because the author has accomplished all of these things. He has written a book that will clearly orient researchers to both the values and the pitfalls of specific methodologies in this complex area of research. At the same time, he has provided a thorough review of accumulated knowledge on lignin biodegradation and transformation. In fact, I know personally that he has rewritten sections of the text numerous times to incorporate newly published information into the book. As such this book will be perhaps the most valuable reference now available to researchers interested in lignin decomposition. It will continue to be an important source of information (especially methodological information) even as rapid research advances continue over the next few years. Indeed, after my first reading of this book, I look at my own lignin degradation research in a new light. I see it less in terms of my own narrow research interests and more in terms of a unique and complex area of study where almost every individual research program is in some way relevant to every other one.

The author deserves considerable credit for so successfully summarizing such a diverse and difficult subject area. As his twin brother and colleague, I hereby become the first to do so in writing.

DON L. CRAWFORD

PREFACE

There is now a tremendous international interest in the subject of lignin biodegradation and transformation. This interest spans many disciplines, including organic chemistry, physical chemistry, plant physiology, plant pathology, ecology, industrial technology, biochemistry, and microbiology. It would probably be impossible for one individual, particularly myself, to adequately discuss all these specialized areas in a book such as this. Thus I have chosen to emphasize the disciplines of biochemistry and microbiology as they concern lignin biodegradation. I have tried, however, to provide enough recent reference materials to lead readers more deeply into other disciplines that I have chosen not to emphasize.

In a research area so diverse as that of lignin biodegradation there are literally hundreds of papers in dozens of journals that concern this topic. I have attempted to discuss as many of the important papers as possible, with an emphasis on the most recent literature. I have unavoidably missed a few significant papers. Please forgive these omissions.

I have also introduced some highly speculative discussions at numerous points in the text. These are intended to encourage imaginative investigation by other scientists. If my speculations are proven inaccurate, so be it. They will have served their purpose in encouraging important research efforts that might otherwise have gone undone.

I would like to thank Don Crawford (University of Idaho) and Kent Kirk (USDA Forest Products Laboratory, Madison, Wisconsin) for helpful comments and discussions during my preparation of this manuscript. I would also like to acknowledge the National Science Foundation (Problem-Focused Research Applications) for support of my research during the past four years.

RONALD L. CRAWFORD

Navarre, Minnesota
February 1981

CONTENTS

Chapter 1

LIGNIN: ECOLOGICAL AND INDUSTRIAL IMPORTANCE

It commonly has been stated that the plant polymer lignin is the second most abundant organic compound on Earth, and that cellulose is the only other organic substance present in the biosphere in larger quantities than lignin (Bellamy, 1974; Nimz, 1974; Crawford and Crawford, 1978; Ander and Eriksson, 1978). Actual quantitative estimates of amounts of lignin and cellulose present in the biosphere are usually omitted from such statements; however, lignocellulosics are biosynthesized by plants in such large quantities that this intuitive ordering of cellulose and lignin as first and second in quantitative importance seems reasonable.

1.1 LIGNIN AS A COMPONENT OF THE BIOSPHERIC CARBON CYCLE

Maintenance of life on Earth is dependent on the continuous operation of many biogeochemical cycles. Quantitatively, one of the most important biospheric cycles is the carbon-oxygen cycle. Solar radiation supplies the driving force that keeps the cycle turning. By trapping the sun's radiant energy by the processes of oxygenogenic photosynthesis, autotrophic organisms convert simple inorganic chemicals into living organic material (biomass) which then flows through the carbon-oxygen cycle. Ultimately living organisms die, and the carbon and oxygen that had been temporarily sequestered is mineralized again to the level of simple inorganic substances. The carbon-oxygen cycle is simple to describe (Figure 1.1), but complex in its operation.

Higher plants comprise one of the unique reservoirs of the carbon-oxygen cycle. These organisms synthesize vast quantities of insoluble, aromatic macromolecules (Evans, 1977), including tannins and, most importantly for

1

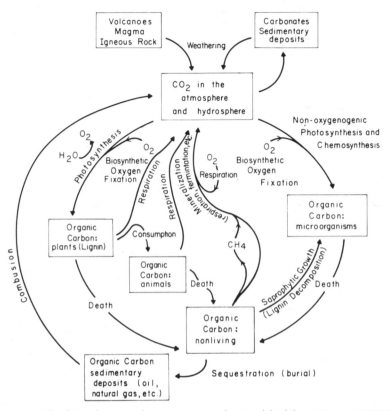

Figure 1.1   The biospheric carbon-oxygen cycle [modified from Evans (1977)].

this discussion, lignins. These aromatic macromolecules are biodegradable, but their biodegradation is often slow. Tannins and lignins are among the most persistent naturally occurring carbon-oxygen complexes. It is probable that the decomposition of tannins and particularly lignins is the rate-limiting step in the biospheric carbon-oxygen cycle (requiring several thousand years for the complete recycling of polyaromatic molecules complexed into soil humic substances). This rate-controlling step of the carbon-oxygen cycle is mediated almost entirely by the catabolic activities of microorganisms.

True lignin (Chapter 2) is distributed widely, but not universally, throughout the plant kingdom. It is found in all *vascular* plants, where it acts as a structural component of support and conducting tissues. Lignin has been unambiguously identified in certain primitive plant groups such as ferns and club mosses, but seems to be absent in the Bryophyta (true mosses) and lower plant groups (Erickson and Miksche, 1974a,b; Miksche and Yasuda, 1978). Nonlignified lower plant groups such as the Bryophyta (Erickson and Miksche, 1974a,b) and even the green algae (Gunnison and Alexander,

1975) often synthesize nonlignin phenolic substances that may falsely analyze as lignin by standard assay procedures such as the Klason technique (Effland, 1977). Thus it is essential that when one examines a plant tissue for the presence or absence of lignin, rigorous chemical characterizations of the polyphenolic material be performed to confirm the presence or absence of lignin's basic monomeric units (Chapter 2; Kratzl and Billek, 1957; Freudenberg, 1968; Erickson and Miksche, 1974a,b).

Higher plants such as angiosperms, gymnosperms, and monocotyledons make up the bulk of organic matter in terrestrial environments and thereby dominate the total biomass of the Earth. The stems of woody angiosperms contain 18–25% lignin on a dry weight basis. Corresponding values for gymnosperms and monocotyledons are 25–35% and 10–30% lignin, respectively (Cowling and Kirk, 1976). Most of the nonlignin plant tissue is cellulosic, with lesser amounts of such things as protein, ash, and extractives (material extractable with solvents such as water, acetone, ethanol, and benzene). Table 1.1 summarizes the general chemical compositions of some representative terrestrial plants.

1.2 LIGNOCELLULOSE IN AGRICULTURE AND INDUSTRY

The biosphere's ability to produce vast quantities of lignocellulosic materials is, of course, the driving force for life on Earth. More practically, however, this capability has been exploited by mankind in the development of large lignocellulose-based industries. Chief among these industries are agriculture, lumbering, and paper making. All these industries produce either directly or indirectly astronomical quantities of lignocellulosic waste materials. These wastes often go largely unused and represent an inexcusable squandering of invaluable natural resources (agricultural residues are valuable to some extent for soil erosion control and soil nutrient replenishment). Estimates of actual amounts of lignocellulosic by-products produced by human technology vary widely. Some recent estimates for the annual production of lignocellulosic wastes in the United States are presented in Table 1.2.

There have been many attempts to develop industrial processes for the conversion of waste lignocellulosics to useful products. Most such attempts have been largely unsuccessful. Lack of success of most attempts, particularly those involving bioconversion (the use of microorganisms as agents to transform wastes to useful products), can be ascribed to the lignin contents of proposed waste substrates. The intimate association of plant polysaccharides with lignin (Sarkanen and Lugwig, 1971) presents accessibility barriers to the removal and utilization of the more valuable polymeric carbohydrates. In fact, as Kirk et al. (1977) point out, most microorganisms

TABLE 1.1   Chemical Composition of Selected Terrestrial Plants

Plant	Constituent, Percent of Dry Weight		
	Holocellulose[a]	Lignin[b]	Other[c]
Birch			
(Betula alleghaniensis)[d]	77.6	19.3	3.1
Spruce			
(Picea excelsa)[d]	70.7	26.3	3.0
Elm			
(Ulmus americana)[e]	74.0	24.0	2.0
Pine			
(Pinus strobus)[e]	68.0	29.0	3.0
Beech			
(Fagus grandifolia)[e]	72.0	22.0	6.0
Hemlock			
(Tsuga canadensis)[e]	64.0	33.0	3.0
Mistletoe			
(Viscum album)[f]	—	22.0	—
Pine			
(Pinus sylvestris)[g]	63.0	28.0	9.0
Red Alder			
(Alnus rubra)[h]	63.1	25.4	11.5
Balsam poplar			
(Populus balsamifera)[h]	66.7	22.3	11.0

[a] Includes cellulose and hemicellulose (glucan, xylan, and mannan).
[b] Klason, or acid-insoluble lignin.
[c] Ash, protein, extractives (components soluble in solvents like acetone, ether, methanol, benzene, and water).
[d] Cowling and Kirk, 1976.
[e] Kirk, 1973b.
[f] Kuroda and Higuchi, 1976.
[g] Ander and Eriksson, 1977.
[h] Eslyn et al., 1975.

that are able to decompose or transform isolated wood polysaccharides cannot attack lignified tissues unless they are also able to degrade the lignin that physically protects the carbohydrates. Alternatively, lignified tissues can be made susceptible to attack by nondegraders of lignin provided the lignin barrier is overcome by physical (e.g., fine grinding of wood) or chemical (e.g., NaOH pretreatment) means. Thus the importance of lignin as the rate-limiting component of the carbon cycle becomes even more apparent since lignin acts as a barrier to slow the decomposition of the even more abundant plant polysaccharides. The importance of lignin as a barrier to microbial degradation of plant tissues is exemplified in the observation that

TABLE 1.2 Annual Production of Lignocellulosic
Wastes in the United States

	10^6 Tons/Year[a]	
Source	Humphrey et al.	Bellamy
Agriculture	400	2347[b-d]
Manure	200	218
Forestry	60	140[e]
Urban	150	256
Industrial	45	110
Miscellaneous		70
Total	925	3141

Source: Bellamy (1974); Humphrey et al. (1977); and
Sitton et al. (1979).
[a] Bellamy's estimates are for 1967. It is not clear for what
year Humphrey et al. produced their estimates.
[b] Includes vegetation wastes normally not collected.
[c] Chahal et al. (1979) estimate the world production of
wheat straw alone to be about 355 million metric tons.
[d] Sitton et al. (1979) quote a value of 299 × 10^6 tons/
year (1977) of agricultural residues in the United States.
[e] Includes mill wastes and residues left in the forest.

lignification is an important mechanism of disease resistance in plants.
Vance et al. (1980) have recently reviewed the literature concerning this
aspect of lignocellulose biodegradation.

Despite the presence of a lignin barrier there is great potential for the use
of microorganisms to biologically transform plant materials and residues into
industrially valuable products. Potential substrates for microbial conversion
include cattle manure (Bellamy, 1972a,b; Bellamy, 1974), urban solid
wastes (Stutzenberger et al., 1970; Stutzenberger, 1971; Updegraff, 1971),
logging and pulpmill wastes (Millett et al., 1970; Mellenberger et al., 1971;
Kirk and Moore, 1972; D.L. Crawford et al., 1973; Harkin et al., 1974;
Pamment et al., 1979), agricultural wastes (Chahal et al., 1979; Chandra and
Jackson, 1971; Dekker and Richards, 1973; Peitersen, 1975; Ramasamy et
al., 1979), and peat (Farnham, 1978; Quierzy et al., 1979; Ghosh and Klass,
1979). Desirable goals for microbial bioconversion of lignocellulosics include bioconversion of industrial lignins to chemicals such as polyphenols
(Goldstein, 1975); bioconversion of lignocellulosics to organic acids (Scott
et al., 1930; Hajny et al., 1951), methane (Schmid, 1975; Clausen et al.,
1977; Hashimoto et al., 1979; Yeck, 1979; Commoner, 1979; Robbins et
al., 1979), glucose (Mandels et al., 1971; Brandt et al., 1972; Dhawan and
Gupta, 1977), alcohols, or single-cell protein (Bellamy, 1974; Daugulis and

Bone, 1977; Moo-Young et al., 1979; Stutzenberger, 1979); microbial delignification (biological pulping) of wood chips (Ander and Eriksson, 1975; Ander and Eriksson, 1978; Eriksson et al., 1976; Kirk and Yang, 1979); microbial pretreatment of highly lignified plant residues to increase their digestibility by ruminants (Dekker and Richards, 1973; Kirk and Moore, 1972; Zadrazil, 1977; Ford, 1978); and simple disposal of troublesome lignocellulosic wastes (Wani and Shinde, 1977).

Efficient and economically viable bioconversion of lignified plant materials is contingent on overcoming the lignin barrier. As the above examples attest, much research has been done with this goal in mind. Numerous efficient lignin-degrading microorganisms are now available for consideration as agents for use in commercial bioconversion processes, although much work remains to be done, particularly in optimizing bioconversion potentials of the best available microbial strains. These microorganisms are discussed in greater detail in Chapter 4.

Chapter 2

THE CHEMICAL STRUCTURE OF LIGNIN AND THE USE OF VARIOUS LIGNIN PREPARATIONS FOR MICROBIOLOGICAL STUDIES

Lignin is among the most complex of biopolymers. The elucidation of its chemical structure has been a formidable task, requiring almost 140 years of effort by some of the world's finest natural products chemists (Nimz, 1974; Kratzl and Billek, 1957). There are still many details of lignin structure to be learned; however, most of the important structural features of lignin probably are now known. As Adler (1977) points out, "Apart from . . . refinements, our present knowledge regarding lignin structure is probably fairly close to the truth, regarding both the qualitative and quantitative aspects."

The importance to microbiologists of understanding the chemical reactions and structural composition of lignin cannot be overemphasized. Lack of such knowledge has been a primary factor slowing progress in the study of lignin biodegradation during much of this century. Real progress in the study of microbial degradation of lignin is quite recent and occurred only after lignin's general structural features became known.

2.1 LIGNIN: BIOSYNTHESIS AND CHEMICAL STRUCTURE

The chemical nature of lignin is known largely from studies of its biosynthesis (Sarkanen and Ludwig, 1971), work pioneered by Freudenberg and his co-workers between about 1930 and 1965. Lignin is an amorphous, three-dimensional, aromatic polymer composed of oxyphenylpropae units. It is formed at the sites of lignification in plants by enzyme-mediated polymerization of three substituted cinnamyl alcohols: p-coumaryl alcohol, coniferyl alcohol, and sinapyl alcohol (Figure 2.1). Phylogenetic origin determines the relative proportions of each alcohol in the lignin of a particular plant (Freudenberg, 1968; Sarkanen and Hergert, 1971).

Higuchi et al. (1977) defined three major groups of lignins: guaiacyl lignin (found in most conifers, lycopods, ferns, and horsetails), guaiacyl-syringyl lignin (found primarily in dicotyledonous angiosperms and a few exceptional gymnosperms), and guaiacyl-syringyl-p-hydroxyphenyl lignin (found in the highly evolved grasses and in compression wood of conifers). Guaiacyl lignin is composed principally of coniferyl alcohol units with small amounts of coumaryl and sinapyl alcohol-derived units. Freudenberg et al. (1962) estimated that the relative proportions of coniferyl, coumaryl, and sinapyl alcohol-derived units in spruce lignin were 80:14:6, respectively, while Erickson and Miksche (1974b) gave somewhat lower estimates than Freudenberg for the amounts of coumaryl and sinapyl units in conifer lignins. Guaiacyl-syringyl lignins contain monomeric units derived from approximately equal amounts of coniferyl alcohol and sinapyl alcohol with only minor amounts of coumaryl alcohol-based units. Freudenberg (1968) stated that lignin in beechwood, representing a typical hardwood, originates from coumaryl, coniferyl, and sinapyl alcohols in a ratio of about 5:49:46. Chang and Sarkanen (1967, 1973), however, have pointed out that hardwood lignins are intrinsically heterogeneous, and that the percentage of syringylpropane units in lignin preparations of four West Coast hardwood species varied from 26% to 60%. Nonetheless, it is clear that hardwood lignins characteristically contain large amounts of sinapyl alcohol-derived monomeric units. Guaiacyl-syringyl-p-hydroxyphenyl lignin is thought to be composed of approximately equal amounts of all three cinnamyl alcohols (Higuchi et al., 1977). However, considerable amounts of p-coumaric acid are bound as esters to grass lignins (Nakamura and Higuchi, 1976; Shimada et al., 1971) and are thus not actually incorporated into the dehydrogenation

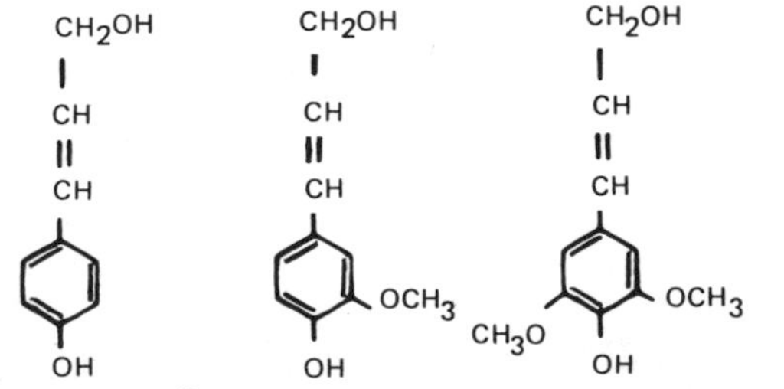

Figure 2.1   The three primary monomeric precursors of lignin.

polymer itself. Adler (1977) has suggested that grass lignins should be classed as normal guaiacyl-syringyl lignins.

Lignin is synthesized by plants from carbon dioxide by way of shikimic acid (Higuchi et al., 1977; Grisebach, 1977). Shikimate is converted to L-phenylalanine, which is the primary precursor of all three of the previously discussed cinnamyl alcohols (Figure 2.1). The cinnamyl alcohols are polymerized enzymatically by phenoloxidases (Chapter 6), enzymes that oxidize the aromatic alcohols, forming free radicals that spontaneously and irreversibly polymerize by complex phenol-coupling reactions (Harkin, 1967). The phenoloxidase involved in lignin biosynthesis probably is a peroxidase (Harkin and Obst, 1973b). The polymerization process results in formation of a highly branched, structurally complex molecule comprised of phenylpropanoid units interconnected by varied types of covalent linkages. This macromolecule is often described as being a dehydrogenative polymerizate (Higuchi et al., 1977).

Figure 2.2 illustrates the biosynthetic pathway whereby plants convert CO_2 to lignin. Several recent reviews give excellent discussions of the enzymes involved in the anabolic sequence (Higuchi et al., 1977; Grisebach, 1977; Hanson and Havir, 1979; Gross, 1977; Gross, 1979), and these are recommended to those readers who wish to delve in depth into the enzymology of lignification.

Figure 2.3 shows a scheme proposed by Adler (1977) for a representative portion of a molecule of conifer lignin and illustrates the types of covalent bonds that have been shown to occur in conifer lignins. The proportions of bond types shown approximate those of natural spruce lignin, though the importance of a few structural types is overemphasized because of the limited number of monomeric units that may be assembled into a schematic of this size (Adler, 1977).

Sakakibara (1980) has recently published his own tentative structural model of softwood lignin. His scheme is based on studies of hydrolysis and hydrogenolysis products from various wood lignins. The Sakakibara model contains most of the important structures shown in Figure 2.3, along with several unusual structural features not reported by other investigators.

Figure 2.4 is a scheme proposed by Nimz (1974) for a representative portion of a molecule of hardwood (beech) lignin. Again, proportions of various bond types, with a few exceptions, quantitatively approximate those that are thought to occur in natural beech lignin. As pointed out by Adler (1977), this schematic presentation probably refers to an "average" lignin, which per se does not exist in the wood.

Apparently there are presently no published schemes that represent a typical grass lignin, though Figure 2.4 may in reality be an appropriate scheme for grass lignins if the absence of esterified cinnamyl alcohols in the figure is taken into account (see Adler, 1977).

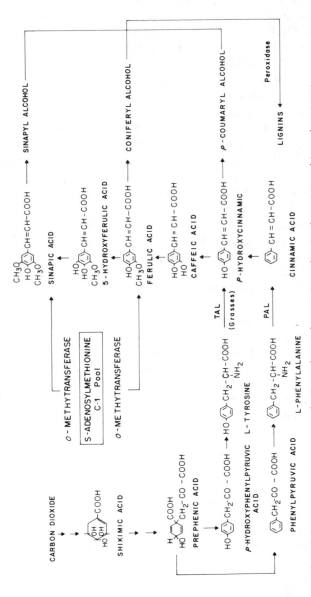

Figure 2.2 Biosynthetic pathway from CO_2 to lignin [from Higuchi et al. (1977)].

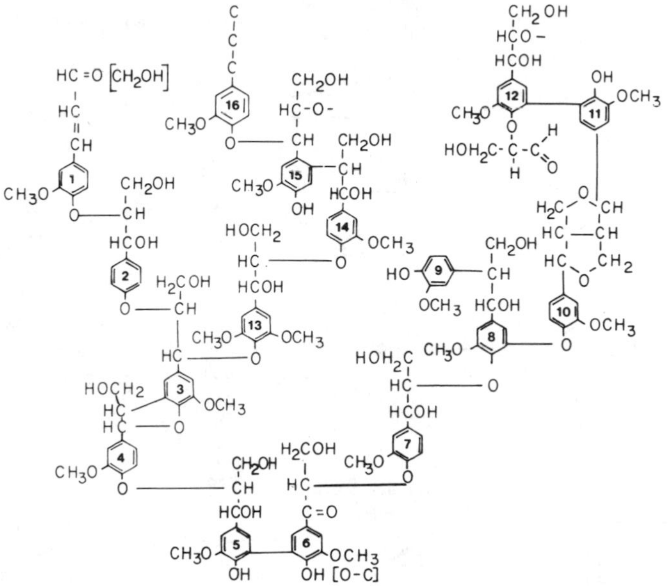

Figure 2.3 Prominent structural features of conifer lignin [from Adler (1977)].

Lignin molecules vary widely in molecular weight, but it is likely that *in vivo* natural lignins have molecular weights in the range of 100,000 daltons or greater (Sarkanen and Ludwig, 1971); thus Figures 2.3 and 2.4 represent only small pieces of total lignin molecules.

The structural schemes shown in Figures 2.3 and 2.4 can be explained readily based on what is known concerning the free-radical mechanism of lignin biogenesis. Obtaining direct analytical proof of such structures is an extremely difficult proposition; however, during the past decades much of this direct proof has been attained. Numerous elegant analytical methods have been developed specifically for the examination of lignin's chemical structure (Nimz, 1974; Adler, 1977). Many of these methods involve extensive depolymerization of lignin using mild chemical procedures that probably do not produce serious side reactions (though a few minor side reactions undoubtedly do occur; see Adler, 1977). These procedures allow isolation of monomeric to tetrameric phenols derived from lignin. Structures of many such phenols have been determined, and have confirmed the general validity of structural schemes such as those of Figures 2.3 and 2.4. Figure 2.5 illustrates a tetrameric phenol recently isolated following mild hydrolysis of lignin. Its structural similarity to the lignin schematics is striking.

Chemical degradation of lignins under strongly oxidizing conditions has also yielded much valuable structural information. Much of this type of structural research has been well summarized by Adler (1977).

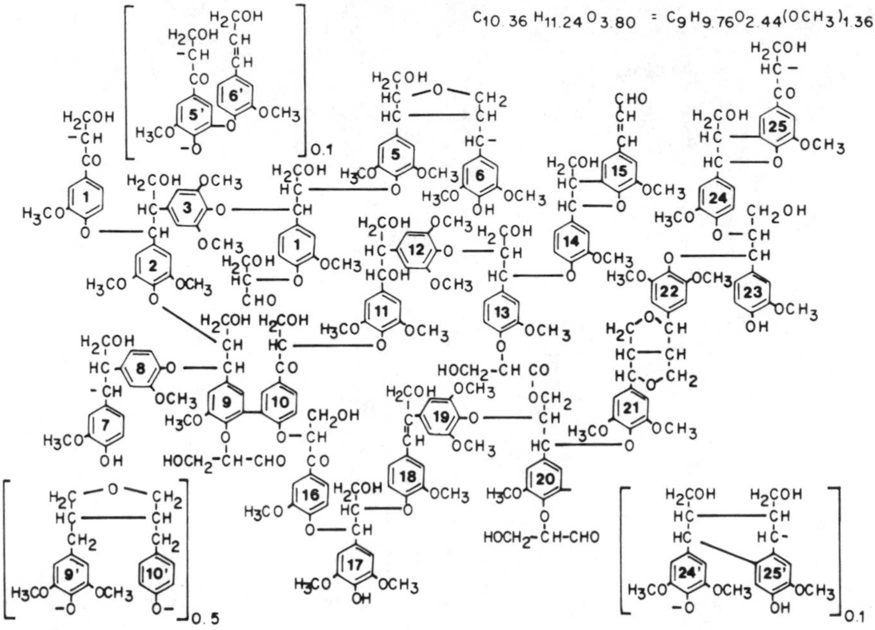

Figure 2.4 A proposed structure for beech lignin [from Nimz (1974); used by permission of Verlag Chemie Inc., *Angewandte Chemie* **86:** 336–344].

Further support for the various structural proposals for lignin has come from the application of ^{1}H-NMR spectroscopy (Lundquist, 1979a,b) and ^{13}C-NMR spectroscopy (Lüdemann and Nimz, 1973; Nimz et al., 1974; Nimz, 1974) to the study of lignin structure. The chemical shifts of nearly all the carbon atoms in the structure shown in Figure 2.4 have been determined from the ^{13}C-NMR spectra of more than 50 lignin model compounds (Nimz, 1977).

Thus Erdtman's proposal in 1933 (Erdtman, 1933) that lignin is formed by dehydrogenative coupling of p-hydroxycinnamyl alcohols has been proven beyond any doubt.

Accumulated data strongly support the existence of at least some chemical bonds between plant polysaccharides and lignin (Lai and Sarkanen, 1971; Kosikova et al., 1979; Hemmingson, 1979). Recent research seems to suggest that galactose and arabinose may act as links between lignins and plant hemicelluloses (Adler, 1977). The effects of such bonds on lignin biodegradation processes are largely unknown.

For in-depth discussions of lignin chemistry and structure, the following references are recommended: Sarkanen and Ludwig (1971); Adler (1968); Nimz (1977); Freudenberg (1968); Harkin (1967); Kirk et al. (1978), and

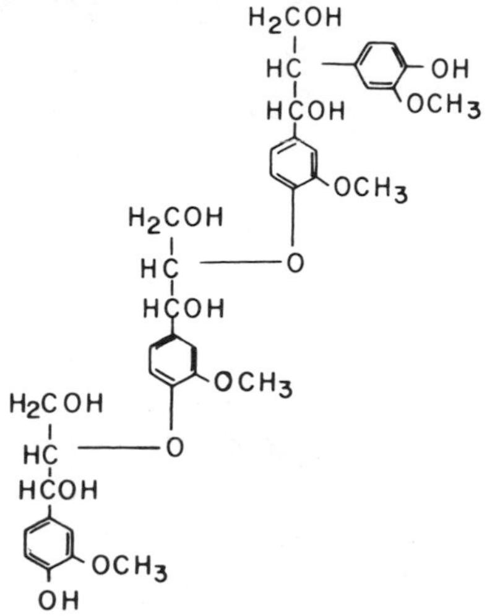

Figure 2.5   Tetrameric phenol isolated following mild hydrolysis of lignin (Nimz, 1974). Isolation of this compound confirms the presence of β-0-4 ether linkages in lignin [see also Lee and Pepper (1978) for the structure of a β-0-4 linked lignin trimer isolated from spruce wood].

particularly Adler (1977). For recent reviews of lignin biosynthesis, see Higuchi et al. (1977), Grisebach (1977), and Hahlbrock and Grisebach (1979).

2.2   LIGNIN AS A SUBSTRATE FOR THE STUDY OF LIGNIN BIODEGRADATION

The preceding discussion of lignin structure emphasizes the fact that lignin is a peculiar biopolymer. Unlike other natural polymers like cellulose, proteins, and nucleic acids, lignin does not have a readily hydrolyzable bond recurring at periodic intervals along a linear backbone. Instead, lignin is a three-dimensional, amorphous polymer with a seemingly random distribution of stable carbon-carbon and ether linkages between monomeric units (Figure 2.6). This type of structure is not amenable to normal modes of biological hydrolysis.

Mechanisms of lignin biodegradation are still largely matters for speculation (Chapter 6); however, it is obvious based on what is known of lignin structure that these biodegradative mechanisms must be unusual and perhaps unique.

BOND TYPE	STRUCTURE	PROPORTION[D] (%)
Arylglycerol-β-aryl ethers	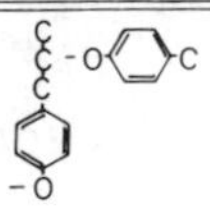	48
Noncyclic benzyl aryl ethers	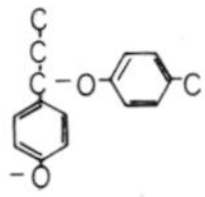	6 − 8
Biphenyl	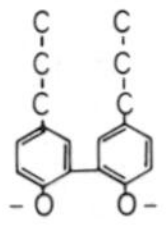	9.5 − 11
1,2 − Diarylpropane structures	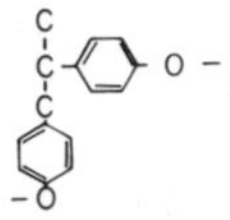	7
Phenylcoumaran structures	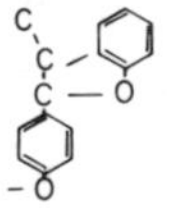	9 − 12
Diphenyl ethers	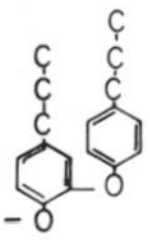	3.5 − 4

Figure 2.6　Proportions of some major types of bonds connecting phenylpropanoid units in spruce (*Picea abies*) lignin (Adler, 1977). The spruce lignin was prepared according to the method of Björkman (1956; 1957a,b; Björkman and Person, 1957a,b). Proportion refers to percent of total phenylpropane units connected to another unit by a particular bond type (some units may have more than one bond type).

2.3　LIGNIN PREPARATIONS FOR USE IN MICROBIOLOGICAL STUDIES

Historically, one of the most serious obstacles to the investigation of lignin biodegradation has been the unavailability of lignin preparations of demonstrated integrity (Kirk, 1971). Many early investigations of lignin biodegradation concerned microbial attack on lignin preparations such as "alkali lignin," "phenol lignin," "HCl lignin," or "Klason lignin." These lignin preparations are highly modified from the natural state, and they should be considered as being of questionable value for microbiological work. For example, Klason lignin (Klason, 1908) is the acid-insoluble material, mostly derived from lignin, remaining following treatment of lignified tissue with cold 72% sulfuric acid, dilution, and refluxing with the diluted acid solution (Pearl, 1967; Effland, 1977). This material is a vastly modified form of lignin, being highly condensed and resinous (Kirk, 1971). It is of no value as a substrate for biodegradation studies (though the Klason procedure for determination of lignin content in plant tissues (Effland, 1977) is still a valuable and widely used technique).

So-called "Brauns' native lignin" is a lignin preparation obtained by extracting ground plant tissues with ethanol. A low-molecular-weight lignin is purified from the ethanol extract (Brauns, 1962). This lignin is obtained in very low yield and is generally considered not to be representative of the bulk of lignin in the extracted tissues (Kirk, 1971), though it may be useful for some biodegradation studies (Cartwright and Holdom, 1973). It should be noted also that some preparations of Brauns' lignin may contain appreciable quantities of lignans, low-molecular-weight, nonlignin contaminants not removed during the purification process (Kratzl and Billek, 1957). On the other hand, Nord and Schubert (1955) published evidence indicating that in some instances (e.g., Scots pine) the greater part of the lignin in wood is chemically uniform, and that Brauns' native lignin is representative of the whole of that lignin [though Hergert (1971) states that IR spectroscopic studies of Brauns' lignin preparations refute this claim]. However, these authors did not have available all the analytical tools and techniques of later investigators, so differences may have gone unnoticed.

In summary, the use of Brauns' native lignin should be discouraged, as there are several better lignin preparations available.

Milled wood lignin, or "Björkman lignin," is a largely unmodified lignin extracted from powdered (ball-milled) plant tissues using neutral solvents without elevated temperature (Björkman, 1956; 1957a,b; Björkman and Person, 1957a,b). Up to 30–50% of the lignin in ball-milled wood can be extracted using 9:1 dioxane–water. The extracted lignin is usually purified by a series of simple solvent precipitations, with occasional addition of a molecular-weight fractionation by gel exclusion chromatography. The average molecular weight of milled wood lignins is 15,000–16,000 (Chang et al., 1975). They typically contain a few percent carbohydrate, but some preparations may have as little as 0.05% carbohydrate contamination (Lundquist and Simonson, 1975). Milled wood lignins are generally believed to be representative of the bulk of lignin in plants (Adler, 1977), except they contain a somewhat increased amount of hydroxyl substituents as a result of the ball-milling process (Chang et al., 1975). They are one of the best lignin preparations for microbiological studies.

Another very good lignin preparation is the so-called cellulase lignin. This lignin is also prepared from ball-milled plant tissues. Milled wood or other plant material is treated with commercially available cellulase—hemicellulase mixtures which remove most of the polysaccharides (Pew and Weyna, 1962). The final product is commonly contaminated by 12–14% carbohydrate. Of all available lignin preparations, cellulase lignin is the most representative of the bulk of a plant's lignin, since cellulase lignin is isolated in an essentially quantitative yield. Though the somewhat high residual carbohydrate content of cellulase lignin may cause some experimental limitations (e.g., microbial growth on carbohydrate rather than lignin), it can be a good preparation for microbiological work.

Another lignin preparation that has now become invaluable in the study of lignin biodegradation is a synthetic lignin (or "model" lignin) generally referred to as DHP (for dehydrogenation polymerizate). As discussed previously (Figure 2.2), the final step in the biosynthesis of lignin is a continuous series of phenol-coupling reactions between peroxidase-generated cinnamyl alcohol radicals. This polymerization reaction can be reproduced *in vitro* using coniferyl alcohol in phosphate buffer containing H_2O_2, peroxidase, and vanillyl alcohol (Sarkanen and Ludwig, 1971; Kirk et al., 1975). Kirk et al. (1975) and Haider and Trojanowski (1975) have published detailed protocols for preparing ^{14}C-labeled DHP's, which have become particularly valuable for microbiological work (Chapter 3). DHP's have been shown by spectroscopic and chemical methods to contain the same types of intermonomer linkages found in natural lignins (Kirk et al., 1975). They are insoluble in water and show number average molecular weights ($\overline{M}_n$) of 1490–1600. Thus DHP's are excellent substrates for the study of lignin biodegradation. They are different from other lignin preparations in that they contain no associated polysaccharides, which may be an advantage under certain experimental situations (Kirk et al., 1976) while a disadvantage in still other situations (Crawford and Crawford, 1976).

Industrially produced lignins are readily available and thus have been used extensively in biodegradation research (Ander and Eriksson, 1978). The two major pulping processes on a worldwide basis are the kraft process and the sulfite process. Millions of tons of kraft lignin and lignin sulfonates are produced annually by the pulping industry. Thus the study of kraft and sulfite lignin degradation by microorganisms is of considerable environmental significance in its own right. There are, however, serious questions as to the relevance of data obtained from the study of industrial lignin biodegradation to the study of biodegradation of natural lignins.

Pulping of plant materials is in its most simple terms the chemical delignification of lignocellulose, usually wood. During the pulping process, lignin is solubilized by degradation and/or derivatization, thereby freeing cellulose fibers for the manufacture of paper and other products (Lundquist et al., 1977). The solubilized lignins are chemically modified forms of natural lignin. Some of the important pulping reactions that modify lignin are

TABLE 2.1 Some Important Pulping Reactions that Modify Lignin

Kraft Process	Acid Sulfite Process
1. Cleavage of α-aryl ether bonds (e.g., between units 4 and 3 of Figure 2.3)	1. Introduction of sulfonic acid groups into α-positions on side chains
2. Cleavage of phenolic β-aryl ether bonds (units 4 and 5 of Figure 2.3) with extensive depolymerization	2. Opening of pinoresinol structures (between units 10 and 11 of Figure 2.3)
3. Limited demethylation of methoxyl groups forming catechol structures	3. Some aryl-alkyl ether cleavages
4. Shortening of some side chains	4. Various condensation reactions, particularly at the α-positions of side chains
5. Various ill-defined condensation reactions	5. Introduction of quinonoid structures
6. Introduction of quinonoid structures	

Source: Gierer (1970).

shown in Table 2.1. It is important, therefore, that investigators evaluate biodegradation data obtained with industrial lignins with the realization that these lignins are significantly modified from the natural state, and that at this point there is insufficient evidence to determine the exact relevance of work with industrial lignins to studies of biodegradation of natural lignins. Lundquist et al. (1977) have published data indicating that chemical changes introduced into lignin by either acid sulfite or kraft pulping processes do significantly alter the manner of lignin degradation by white-rot fungi (Chapter 4). Trojanowski and Leonowicz (1969) and Hiroi and Eriksson (1976) have shown that kraft lignin is degraded more rapidly than milled wood lignin by Pleurotus ostreatus, and that kraft lignin degradation is less cellulose dependent than milled wood lignin degradation by this fungus [this latter observation has been confirmed by Ander and Eriksson (1977)]. On the other hand, a large volume of work by Eriksson and co-workers (Ander and Eriksson, 1978) indicates that studies of the fungal degradation of kraft lignin are probably relevant to the study of degradation of lignin in wood. This question of relevance concerning degradation of industrial lignins as compared to natural lignins is important and is deserving of study as a question in its own right.

As Lundquist and Kirk (1980) point out, kraft lignins have often been studied with little if any purification (Deschamps et al., 1980). In such instances the kraft lignin preparations (e.g., Indulin ATR-C from Westvaco Corp., North Charleston, S.C.) usually contain nonlignin-derived materials and low-molecular-weight kraft lignin degradation products. Thus it is usually very difficult to interpret microbiological research data obtained using unpurified kraft lignins. Lundquist and Kirk (1980) have described a simple purification procedure that removes nonlignin contaminants (up to 40% of the total weight) from industrial kraft lignin preparations. This or some similar procedure should be used to purify industrial kraft lignins prior to their use in microbiological experiments.

Certain Basidiomycetes known as brown-rot fungi (Chapter 4) decay wood and leave a residue approximately equal to the amount of lignin in the original wood (Kirk and Adler, 1970). This residue has been called "enzymatically liberated lignin" (Schubert and Nord, 1950), and has been used by some investigators as a substrate for biodegradation studies. Brown et al. (1968) summarized the published literature concerning the chemical and physical characteristics of brown-rotted lignin, finding good evidence for the following: (a) the distribution of lignin in brown-rotted wood cells is substantially the same as in the sound wood prior to decay; (b) the amount of fungal mycelium remaining in brown-rotted wood at any stage of decay is a negligible part of the total weight remaining; (c) the brown color of this lignin is indicative of fungus-mediated chemical modifications to the lignin,

TABLE 2.2   Usefulness of Various Lignin Preparations for Microbiological Growth and Transformation Studies

Lignin Preparation	Usefulness
Klason lignin	[a]
Kraft lignin	[b]
Lignin sulfonates	[b]
Enzymatically liberated lignin	[b]
Acidolysis lignin	[b]
Brauns' native lignin	[c]
Cellulase lignin	[d]
DHP lignin	[d]
Milled wood lignin	[e]

[a] Should not be used (also included in this category are various other lignin preparations such as thioglycollate-lignin, HCl-lignin, and phenol-lignin).
[b] Useful only under rigidly defined circumstances where chemical modifications of the lignin molecule do not interfere with interpretation of results (e.g., as a carbon source for isolation of microorganisms).
[c] Useful for many applications.
[d] Good for general use.
[e] Probably the best preparation available.

but the nature of these changes is not well understood; and (d) the methoxyl content of the decayed lignin is reduced by as much as 25% [confirmed by Kirk and Adler (1970)]. Brown-rotted lignin may contain about twice as many α-carbonyl groups on its propane side chains as compared to sound lignin (Kirk, 1975). Thus enzymatically liberated lignins are sufficiently chemically modified that care should be exercised in using them for microbiological studies. They are, however, potentially as usable for such work as are kraft lignins.

A final lignin preparation sometimes used for microbiological studies is "acidolysis lignin" (Odier and Monties, 1977). This lignin is also referred to as "dioxane lignin" (Odier and Monties, 1978a,b). Acidolysis lignin is usually solubilized from extractive-free plant tissues by refluxing the tissues with dioxane–water (9:1 v/v) containing the equivalent of 0.2M HCl. This procedure has been particularly valuable for studies of lignin structure (Adler, 1977). A high-molecular-weight lignin is solubilized by acidolysis and then purified by precipitation in ether and/or water (both being preferred). The acidolysis procedure is known to modify numerous lignin substructures, including β-ether linkages, propanoid side chains, α-aryl ether linkages, phenylcoumaran moieties, 1,2-diarylpropane-1,3-diol moieties, and glyceraldehyde-2-aryl ether structures (Adler, 1977). Undoubtedly many other degradative reactions also occur during acidolysis. Thus dioxane lignins are useful for microbiological work with at least the same restrictions that apply to other chemically modified lignins (Table 2.2).

Chapter 3

METHODS FOR THE STUDY OF LIGNIN BIODEGRADATION

As discussed in the preceding chapter, lack of knowledge of the chemical structure of lignins has been a major obstacle to the study of lignin biodegradation. This obstacle has now been overcome to a large extent. Another serious obstacle to the study of lignin biodegradation during the past 50 years has been inadequacies in available methodologies for application to this difficult research problem (Crawford and Crawford, 1976; 1978). Serious methodological inadequacies still exist; however, the past decade has seen much progress toward resolution of many of the most troubling technical problems. Much of this progress is derived from the application of radioisotopic methods to the study of lignin biodegradation. These methods utilize ^{14}C as a tag for following the carbon atoms of lignin as the polymer undergoes biological transformations. In the following sections I discuss what may be termed "classical" methods for the study of lignin biodegradation. I then examine the recently introduced radioisotopic procedures and their application to lignin biodegradation research. Finally, I discuss the use of lignin model compounds as a means to study lignin biodegradation.

3.1 CLASSICAL METHODS FOR THE STUDY OF LIGNIN BIODEGRADATION

Probably the most useful of the nonisotopic techniques for the study of lignin biodegradation has been the so-called soil block procedure. In this method extractive-free lignocelluloses (usually wood wafers) are placed on the surface of sterile soil or vermiculite contained within a chamber where humidity and temperature are controlled within narrow limits (ASTM, 1962; Ander and Eriksson, 1977: Kirk and Moore, 1972). A large number of replicate samples are inoculated with a microorganism (usually a fungus) and incubated under defined conditions, often for several months. Periodically samples are removed for weight loss determinations and chemical analyses. Samples typically are scraped free of fungal mycelium, weighed, milled to pass through a 40-mesh sieve, and then analyzed for Klason lignin (Effland, 1977), total reducing sugars (Moore and Johnson, 1967), and/or holocellulose (Kirk, 1973), mannan, glucan, and xylan (Ander and Eriksson, 1977; Ander and Eriksson, 1975a,b; Sjöstrom et al., 1966; Kirk and Highley, 1973). These procedures generally allow reliable estimates of rates and extents of biodegradation of lignin and carbohydrates in lignocellulosic materials, particularly wood. Major disadvantages of the various soil block procedures include (a) its time consumption, (b) the fact that the Klason procedure for quantitation of lignin often gives only a rough estimate of the true lignin content of numerous plant tissues (particularly for sound wood of hardwoods or microbiologically transformed tissues; see Crawford et al., 1979, and Kirk, 1975), and (c) the unavoidable development of deterioration gradients and zones within the lignocellulose sample being decayed (Beall et al., 1976).

During investigations of microbial degradation of lignin preparations such as Brauns' native lignin, milled wood lignin, DHP's, acidolysis lignin, and kraft lignin a common technique has been to estimate lignin decomposition rates and extents by observing decreases in absorbance of lignin preparations at 280 nm. For example, Odier and Monties (1977; 1978a,b) used this technique to study the bacterial degradation of acidolysis lignin prepared from wheat straw. These authors used a value of $\epsilon_{1cm}^{1\%}$ (280 nm) = 192.5 for lignin dissolved in 1 : 1 dioxane–water. Hiroi and Eriksson (1976) quantified residual lignin by dissolving it in acetyl bromide–acetic acid (specific absorbancy, 280 nm = 20.9 l/g-cm) or $2M$ NaOH (specific absorbancy, 280 nm = 23.0 l/g-cm). Iwahara et al. (1977) estimated extents of microbial degradation of coniferyl alcohol DHP by observing the percent OD decrement at 280 nm. These approaches can be very valuable if they are used with caution; however, there are some problems with spectroscopic quantification methods. At the termination of a microbial incubation, residual lignin must be reisolated from the culture media. This is usually accomplished by centrifuging or filtering off insoluble materials (cell mass plus residual lignin) and dissolving lignin from this residue using dioxane–water (Odier and Monties, 1977), acetyl bromide (Hiroi and Eriksson, 1976), or dilute NaOH (Ferm and Nilsson, 1969). The most serious difficulty encountered with such procedures has been adsorption between lignin and microbial cell surfaces (Gottlieb and Pelczar, 1951). It is difficult to establish that all adsorbed lignin is extractable with dioxane–water or NaOH solutions. This is much less of a problem with acetyl bromide, which usually dissolves the entire residue

(Johnson et al., 1961). Improper accounting of adsorption phenomena may result in artifactually high estimates of lignin degraded.

Another potential problem with spectrophotometric determinations of lignin remaining in culture media involves possible changes in $\epsilon_{(280)}$ resulting from microbial modifications of lignin's chemical structure (Hiroi and Eriksson, 1976). For example, the UV absorption spectrum of spruce lignin modified by the brown-rot fungus Lenzites trabea (=Gloeophyllum trabeum) is quite different from that of sound lignin (Kirk, 1975), though lignin from wood decayed by the brown-rot fungi Lentinus lepideus, Poria vaillanti, and Lenzites sepiaria showed UV/visible absorption spectra identical to that of nondecayed lignin (Nord and Schubert, 1955). A comparison of sound and white-rotted (Polyporus versicolor [=Coriolus versicolor]) lignins from sweetgum showed that the UV spectra were essentially the same (Kirk and Lundquist, 1970), while other white-rotted lignins (e.g., spruce lignin decayed by Coriolus versicolor or Polyporus anceps) yielded UV spectra that had markedly increased absorptivities at about 260 and 300 nm as compared to nondecayed lignins (Kirk and Chang, 1974). These observations point out that, though it may be true in particular instances, it is not safe to assume that ϵ_{280} is unchanged from that of sound lignin by microbial modifications of residual lignin.

Analysis of the methoxyl content of lignocellulose has been used as a means to estimate lignin depletion during microbial decay of plant tissues (Trojanowski et al., 1977; Jaschhof, 1964). This method does not distinguish between extensive lignin degradation and simple demethylation of lignin (Trojanowski et al., 1977), and is therefore unreliable as an estimate of lignin degradation per se.

A procedure based on chlorine consumption has been used to follow lignin degradation rates (Ander and Eriksson, 1977; Hiroi and Eriksson, 1976). Chlorine number is apparently a linear function of plant tissue lignin content, up to at least a total of 20 mg of lignin per sample (Hiroi and Eriksson, 1976). Chlorine numbers, however, differ for different lignin preparations, reflecting the dependence of the procedure on the chemical structure of lignin (Kryklund and Strandell, 1967). Hiroi and Eriksson (1976) showed that the chlorine numbers of lignosulfonate decayed by Pleurotus ostreatus are 10–15% lower than those of the original lignosulfonate preparation (results indicative of fungal-mediated structural changes in the lignin polymer which depress chlorine consumption). Thus use of the chlorine number procedure in some instances may result in estimates of lignin losses that are actually higher than true losses. It is probable that decay of lignin by a different organism (mediating different structural changes to the lignin polymer) might result in either no change in chlorine number, or even an increase in chlorine number as compared to nondecayed lignin. The chlorine number procedure is, in addition to the above disadvantages, sensitive to the presence of phosphate (a common growth media component). For these reasons the chlorine number method cannot be recommended as a generally useful technique for estimating lignin losses during decay of lignocellulosics.

Another chemical method for lignin determination that shows promise for degradation studies is the acetylbromide method of Johnson et al. (1961). However, this method is a spectroscopic procedure (based on ϵ_{280} of acetylbromide-solubilized plant tissue), and suffers the disadvantages already discussed for other UV-absorption methods. Recently vanZyl (1978) found that the acetylbromide procedure yielded lignin values showing variations of up to 10%. Various hard to control parameters were found to affect the reproducibility of the method.

Sundman and Näse (1971) developed a simple plate test for direct visualization of microbiological lignin decomposition. In this procedure an agar medium containing lignin and another energy source (e.g., glucose) is inoculated with a test organism. After incubation under appropriate conditions, the plates are stained by flooding them with a solution of ferric chloride–potassium ferricyanide in water. Clear zones around or beneath colonies indicate lignin degradation. This procedure is qualitative and is not applicable to quantitative studies. However, it is adaptable for use with most lignin preparations used in microbiological work (Table 2.2), and is valuable as a screening technique.

Microbial growth (increases in cell mass and/or cell number) at the expense of lignin has been used by some investigators to establish the capability of pure microbial cultures to degrade lignin. For example, Iwahara et al. (1977) grew various molds in a defined medium containing coniferyl alcohol DHP as the sole carbon source, and determined the yield of mycelium formed at the expense of DHP carbon. Odier and Monties (1978) grew a Xanthomonas species in a defined medium containing acidolysis lignin as the sole carbon source, and followed the increase in bacterial cell number as the organism multiplied at the expense of lignin carbon. Such experiments are valuable as a means to provide supplemental evidence for an organism's ability to degrade lignin. However, the following experimental problems must be taken into account to allow the drawing of valid conclusions from such microbial growth experiments: (a) the lignin preparations used must be free of nonlignin contaminants (e.g., carbohydrates) that might provide another source of carbon for microbial growth; (b) lignin should not be added to a culture medium as a solution in a solvent (e.g., ethanol, acetone) that might serve as a carbon source for microbial growth; and (c) increases in numbers of bacteria in a growth medium containing lignin as a sole carbon source should be in excess of increases that might result from cell division at the expense of preexisting energy or carbon pools.

Measurements of microbial respiration rates in the presence and absence of lignin have been used as indicators of an organism's ability to oxidize lignin (see Odier and Monties, 1978a). The same types of experimental difficulties apply to this technique as were discussed earlier for microbial growth experiments. In particular, oxidation of nonlignin contaminants and/or lignin solvents must be taken into account. As pointed out by Gottlieb and Pelczar (1951), use of oxygen absorption and bacterial numbers is a criterion of growth, and does not unequivocally indicate lignin utilization unless experiments are particularly well controlled.

Direct microscopic observation of lignocellulose decomposition, particularly observations using scanning and transmission electron microscopy, can be of considerable value in investigations of decay processes. For example, Sutherland et al. (1979) used scanning electron microscopy very successfully to show colonization of sterile Douglas fir (Pseudotsuga menziesii) phloem by the actinomycete Streptomyces flavovirens. Microscopic observations indicated that the nonlignified walls of parenchyma cells were attacked first, followed by attack on the thick-walled, heavily lignified sclereids. Figure 3.1 shows scanning electron micrographs of Douglas fir phloem before and after decay by Streptomyces flavovirens.

Schmidt (1978) used an optical microscope to observe decay of beechwood sections before and after partial delignification. A Cellulomonas sp. and Bacillus polymyxa were shown to decompose partially delignified cell walls but not fully lignified walls. Bacterial attack on delignified walls commenced at the S_1 layer and then advanced into the S_2 layer. Highly lignified parts of the walls resisted bacterial attack. It would seem that a combination of techniques such as those used by Schmidt (1978) and the UV microscopy procedures developed by Fergus et al. (1969) for measuring the distribution of lignin across plant cell walls could yield some unique information concerning the ultrastructure of lignin decomposition by lignocellulose-degrading microorganisms.

3.2 RADIOISOTOPIC METHODOLOGY AND THE STUDY OF LIGNIN BIODEGRADATION

In recent years there has been a surge of progress in the study of lignin biodegradation, largely resulting from the application of radioisotopic assay techniques to the problem (Crawford and Crawford, 1978). These techniques have alleviated many of the experimental difficulties discussed in the preceding section. Radioisotopic methods for the study of lignin degradation utilize ^{14}C as a tag to follow the fate of lignin carbon atoms as they undergo biotransformation processes. These methods have allowed for the first time unequivocal and highly sensitive assays of lignin biodegradation.

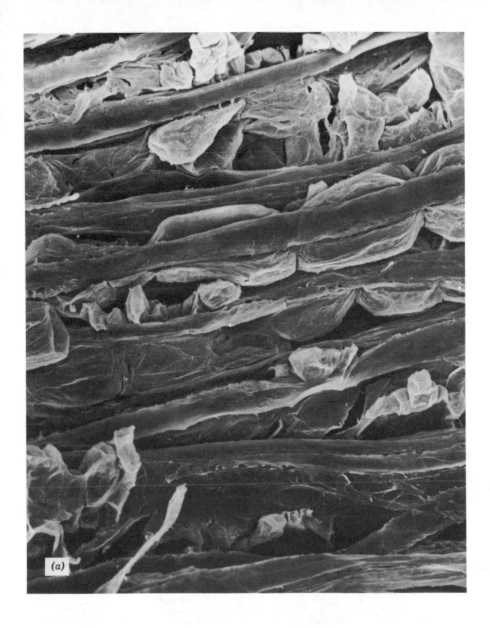

Figure 3.1 Colonization and degradation of Douglas fir phloem by *Streptomyces flavovirens* [scanning electron micrographs provided by R. A. Blanchette; from Sutherland *et al*. (1979)]. (A) Noninoculated phloem tissue (approx. 560×) showing intact cell walls. (B) Decay after 2 weeks (approx. 2,500×) showing zones of disruption in parenchyma cells. (C) Decay after 5 weeks (approx. 12,000×) showing extensive mycelial growth and tissue decay (sclerid decay was not observed until about 9 weeks postinoculation)

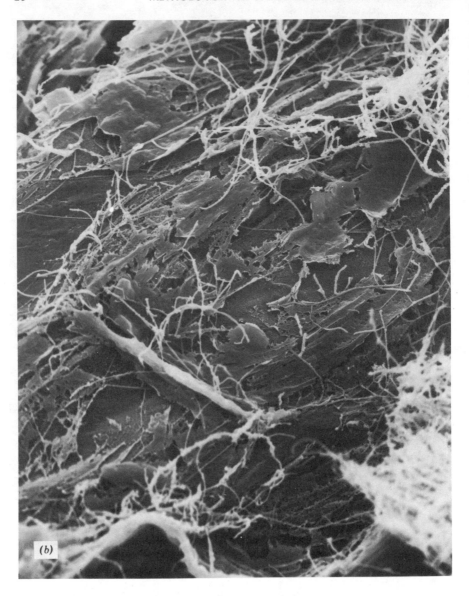

(b)

Radioisotopic assay techniques have been applied successfully to a wide range of lignin-related research problems (Crawford and Crawford, 1978), including (a) determination of the true range of microbial taxa that are able to decompose lignin (Chapter 4), (b) screening of microbes for lignin-decomposing ability (Haider *et al.*, 1978), (c) kinetics of lignin degradation in

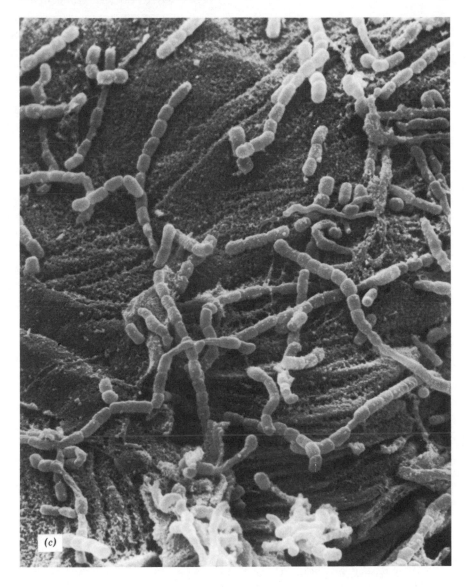

natural environments (Hackett et al., 1977), (d) enumeration of lignin-degrading microorganisms (R. L. Crawford et al., 1977), (e) examination of the biochemistry and physiology of lignin degradation by white-rot fungi (Chapter 6), and (f) examination of the biodegradability of industrial lignins (Lundquist et al., 1977).

Radioisotopic procedures for the study of lignin biodegradation are based on the use of ^{14}C-labeled lignin preparations. Preparations include ^{14}C-DHP's (synthetic lignins), ^{14}C-milled wood lignins, and [14]C-[LIGNIN]-lignocelluloses (natural lignocelluloses labeled specifically in their lignin components). Biodegradative processes result in the conversion of ^{14}C-lignins into [14]CO_2 (or other ^{14}C-labeled degradation products) that can be trapped and quantified. In this way information may be obtained concerning both rates and extents of lignin biodegradation under given sets of experimental conditions.

In the following section I discuss the preparation, characterization, use, advantages, and limitations of the various ^{14}C-labeled lignins now used for microbiological research. Data accumulated by the various laboratories using radioisotopic procedures are largely presented in later chapters.

3.2.1   Preparation of ^{14}C-Labeled Natural Lignocelluloses

Lignin is biosynthesized in vascular plants by a branching sequence of reactions, as shown in Figure 2.2. This anabolic sequence was elucidated largely by observing dilution of radioactivity during incorporation of [14]C- or ^{3}H-labeled compounds into the lignins of growing plants (Brown and Neish, 1955; Eberhardt and Schubert, 1956; Pickett-Heaps, 1968; Terashima et al., 1975; Hasegawa et al., 1960; Freudenberg, 1956; Kratzl and Billek, 1957). It has been possible to use these labeling techniques not only to examine lignin biosynthesis, but also to study lignin biodegradation (Crawford and Crawford, 1978; R. L. Crawford et al., 1979b). This was accomplished by feeding plants ^{14}C-labeled precursors such as ^{14}C-phenylalanine (Crawford and Crawford, 1976), ^{14}C-ferulic acid (Haider and Trojanowski, 1975), ^{14}C-cinnamic acid (Phelan et al., 1979), or ^{14}C-p-coumaric acid (Haider et al., 1977; Phelan et al., 1979). These precursors are usually incorporated preferentially into a plant's lignin, and with the proper work-up procedures to remove low-molecular-weight ^{14}C-labeled contaminants, [14]C-[LIGNIN]-lignocelluloses can be prepared, that is, lignocelluloses containing ^{14}C in their lignin components, but not in other tissue fractions.

Procedures for producing [14]C-[LIGNIN]-lignocelluloses are relatively simple. A typical procedure using the cut stem method for feeding plants ^{14}C-labeled lignin precursors is summarized in Table 3.1. Injury to a plant (e.g., by cutting a stem) is known to induce physiological changes in plant metabolism, possibly affecting processes of lignin biosynthesis. Thus other methods of administering ^{14}C-labeled lignin precursors may be preferred by some investigators. Some representative administration procedures are listed in Table 3.2.

Work-up may also vary somewhat from the procedures suggested in Table 3.1. For example, Haider el al. (1977) used the following steps to remove

TABLE 3.1   Preparation of ¹⁴C-[LIGNIN]-lignocelluloses by the Cut Stem Procedure

Step	Protocol
1.	Remove a small limb or stalk from the plant of interest and prepare the cut surface (under water) so that it is as nearly flat as possible.
2.	Immerse the cut surface in a small volume of dilute buffer (pH 7) containing a ¹⁴C-labeled lignin precursor of high specific radioactivity (in a typical experiment 10–50 µCi of labeled precursor is used).
3.	Allow the plant to absorb the buffer solution; do not allow the solution container to become dry.[a]
4.	Allow the plant to metabolize at room temperature for 1–7 days.
	(a) Continue this step until the plant wilts.
	(b) This should be done in a hood or greenhouse.
5.	Remove the sapwood (bark removed, if possible) from tree specimens or use the intact stalk for grass specimens.[a]
	(a) Dry the material at 60°C.
	(b) Grind it to pass a 40-mesh sieve.
	(c) Use appropriate precautions to prevent contamination of work areas by radioactive plant particles.
6.	Treat the milled plant material to the following series of solvent extractions:
	(a) Water at 60–80°C for 4 h; repeated once.
	(b) Benzene–ethanol (1:1) Soxhlet extraction for 8 h (removes extractives and Brauns' lignin fraction).
	(c) Absolute ethanol Soxhlet extraction for 4 h; repeat until the ethanol extract is visually colorless.
	(d) Water at 60–80°C until no additional significant radioactivity is removed.
	(e) See Kratzl and Billek (1957) for a discussion of extraction efficiencies.
7.	Dry the plant material at 60°C; store in properly labeled containers (a desiccator is recommended).
8.	Determine by appropriate methods (see text) the distribution of radioactivity within the plant's various structural components.

Source: Crawford and Crawford (1978).

[a]Barnes and Friend (1975) have reported that 80% of the ¹⁴C-phenylalanine and ¹⁴C-cinnamate they administered to detached leaves of representative dicots and monocots remains in the basal portion of the leaves. Thus a product of higher specific activity might be obtained by the cut stem procedure if only the basal portions of stems were harvested. No systematic study has been performed to examine this possibility for most plants; however, we have observed that peripheral Douglas fir needles become radioactive when radioactive lignin precursors are administered to fir stems (Crawford and Crawford, unpublished observations).

extractives from their ¹⁴C-[LIGNIN]-lignocelluloses: (a) after harvest, grind the plant tissues in liquid nitrogen; (b) freeze-dry the material; (c) extract the freeze-dried powder 3 times with 80% boiling ethanol; and (d) dry the extracted material prior to use in microbiological experiments. Trojanowski

TABLE 3.2　Methods for Administering ^{14}C-Labeled Lignin Precursors to Plants

Plant	^{14}C-Labeled Compound	Administration Method	Reference
Zea mays L. (sprouts)	p-Coumaric acid	Injection	Haider et al. (1977)
Zea mays L. (stalks)	Ferulic acid	Injection	Haider and Trojanowski (1975)
Zea mays L. (seedlings)	Ferulic acid	Absorption	Trojanowski et al. (1977)
Viscum album (shoots)	Phenylalanine	Absorption	Kuroda and Higuchi (1976)
Populus trichocarpa	Ferulic acid	Cut stems	Saleh et al. (1967)
Triticum vulgare	Shikimic acid	Cut stems	Brown and Neish (1955)
Saccharum officinarum	Shikimic acid	Cut leaf tips	Eberhardt and Schubert (1956)
Triticum vulgare (shoots)	Various compounds	Absorption	Pickett-Heaps (1968)
Populus nigra (hybrid)	Various compounds	Cut shoots	Terashima et al. (1975)
Pinus strobus (tissue culture)	Various compounds	Absorption	Hasegawa et al. (1960)
Grasses	Various compounds	Internodal injection	Kratzl and Billek (1957)
Trees	Various compounds	Implantation	Kratzl and Billek (1957)
Populus tremuloides	Phenylalanine	Cut stems	Reid (1979)
Pseudotsuga menziesii	Cinnamic acid and p-coumaric acid	Cut stems	Phelan et al. (1979)

et al. (1977) followed approximately the same procedure with the addition of an extraction with cold dilute NaOH. The most important differences between these work-up procedures and the procedure outlined in Table 3.2 are (a) the omission of a benzene extraction and, in the latter instance, (b) addition of an alkaline H_2O wash. Omission of a benzene extraction for some lignocelluloses can be expected to leave in the plant tissue some nonpolar extractives that are removed by the procedure of Table 3.1. The NaOH treatment (though mild) slightly delignifies the lignocellulosic substrate, and the final lignocellulose product is somewhat more susceptible to microbial attack than similar material not subjected to NaOH pretreatment (Kirk *et al.*, 1977). However, it can be noted that even extraction of plant tissues with hot distilled water (Table 3.1, step 6a) removes some lignin (Kratzl, 1950). It appears necessary to accept these small lignin losses (along with loss of the Brauns' lignin component) to ensure that nonlignin ^{14}C-labeled materials are removed from the final product.

3.2.2 Chemical Characterization of ^{14}C-[LIGNIN]-lignocelluloses

The major problem associated with ^{14}C-[LIGNIN]-lignocelluloses is the difficulty of characterizing them in concise, chemical terms (Reid, 1979). Numerous, often inexact chemical analyses must be performed on these ^{14}C-lignocelluloses to establish the distribution of ^{14}C within the lignocellulose complex. Experience has taught that it is advisable to chemically characterize every fresh preparation of ^{14}C-lignocellulose by as many analytical procedures as are available. This is particularly true if one uses this labeling technique with a plant species that has not been previously examined for its labeling pattern following administration of ^{14}C-labeled lignin precursors (R. L. Crawford *et al.*, 1979b). At least the following analyses should be performed: (a) Klason analysis for acid-soluble and acid-insoluble ^{14}C; (b) chromatographic separation of acid-hydrolyzable wood sugars and aromatic amino acids to confirm absence of ^{14}C in these plant components; and (c) treatment of ^{14}C-lignocelluloses with NaOH (1N, 25°C for 24 h) to estimate the amount of ^{14}C-labeled compounds esterified to plant lignins (see Shimada *et al.*, 1971; Nakamura and Higuchi, 1976).

^{14}C-[LIGNIN]-lignocelluloses have been prepared from at least the following plants: fir (*Pseudotsuga menziesii*), American elm (*Ulmus americanus*), eastern hemlock (*Tsuga canadensis*), Virginia pine (*Pinus virginiana*), eastern red cedar (*Juniperus virginiana*), white oak (*Quercus alba*), red maple (*Acer rubrum*), black gum (*Nyssa sylvatica*), cattail (*Typha latifolia*), and maize (*Zea mays* L.) (R. L. Crawford *et al.*, 1979b). Table 3.3 summarizes the distribution of acid-soluble and acid-insoluble ^{14}C in specific preparations of these various ^{14}C-lignocelluloses, as determined by the Klason procedure (Effland, 1977). This table shows that in all cases a higher percentage of ^{14}C was localized in the Klason lignin fraction than in the acid-soluble fraction of lignocelluloses labeled by feeding plants ^{14}C-phenylalanine, ^{14}C-ferulate, or ^{14}C-coumarate. It should be realized, however, that the Klason procedure for lignin analysis often gives only a rough approximation of the distribution of ^{14}C between polysaccharides, proteins, and lignin (R. L. Crawford *et al.*, 1979b). The Klason procedure has a serious drawback in that considerable

TABLE 3.3 Distribution of ^{14}C in Various ^{14}C-[LIGNIN]-lignocelluloses as Determined by the Klason Acid-Hydrolysis Procedure

Plant	dpm/mg	% dpm in Klason Lignin	% dpm Acid Soluble	% dpm Recovered
Hemlock (*Tsuga canadensis*)[a]	6500	52	36	88
Virginia pine (*Pinus virginiana*)[a]	4400	56	35	91
Red cedar (*Juniperus virginiana*)[a]	6000	42	32	74
Red cedar (*Juniperus virginiana*)[a]	5053	95	13	107
White Oak (*Quercus alba*)[a]	6900	45	40	85
Red maple (*Acer rubrum*)[a]	1900	70	26	96
Red maple (*Acer rubrum*)[a]	1313	44	34	78
Black gum (*Nyssa sylvatica*)[a]	6000	48	46	94
Cattail (*Typha latifolia*)[a]	3000	93	2	95
Cattail (*Typha latifolia*)[a]	1433	73	15	88
Spruce (*Picea excelsa = P. abies*)[a]	438	65	46	111
Spruce (*Picea excelsa*)[b]	390	53	47	100
Douglas fir (*Pseudotsuga menziesii*)[a]	1654	61	32	93
Elm (*Ulmus americana*)[b]	603	63	35	98
Fir (*Pseudotsuga menziesii*)[c]	105	84	15	99
White oak (*Quercus alba*)[a]	706	80	20	100
Aspen (*Populus tremuloides*)[a,d]	1500	—	17	—

[a] Labeled by feeding ^{14}C-[U]-phenylalanine.
[b] Labeled by feeding ^{14}C-[side chain]-ferulic acid.
[c] Labeled by feeding ^{14}C-[ring]-*p*-courmaric acid.
[d] Reid (1979).

72% H_2SO_4–soluble lignin is present in many lignocelluloses (Migata and Kawamura, 1944). The values observed for acid-soluble ^{14}C are often high because of solubilization of acid-soluble lignin. Some of the acid-soluble ^{14}C may be derived from ^{14}C-labeled phenolics that are esterified to newly synthesized lignins; however, our experiments indicate that this is usually not a serious problem (Table 3.4). It is probable that some ^{14}C is incorporated into peripheral units of lignin which are more susceptible to acid hydrolysis than are more highly condensed lignins.

Additional analytical data are necessary to make more specific conclusions as to the distribution of ^{14}C in the wood components of ^{14}C-lignocelluloses. For example, numerous [14]C-[LIGNIN]-lignocelluloses (tree lignocelluloses in particular) have been examined for contamination of polysaccharides and/or proteins by [14]C. Little radioactivity can be found associated with wood sugars and/or aromatic amino acids when these are isolated by paper or thin-layer chromatography of wood hydrolysates (R. L. Crawford et al., 1979b; D. L. Crawford, 1978). Incorporation of ^{14}C-phenylalanine into plant proteins can pose a problem with certain plant species, particularly plants of high protein content. This problem, if present, is revealed during examination of isolated amino acids. If labeling of proteins is minor (the usual case; e.g., Crawford, 1978), minimum values for [14]CO$_2$ recovery from [14]C-[LIGNIN]-lignocelluloses may be established (e.g., the 2% proposed by Crawford and Crawford, 1976) before ascribing lignin-degrading ability to any particular microbial incubation (Crawford et al., 1979). Serious protein contamination problems can be overcome either by

TABLE 3.4   Solubilization of ^{14}C from Various [14]C-[LIGNIN]-lignocelluloses by Treatment with 1N NaOH

[14] C-Labeled Precursor[a]	% ^{14}C-Solubilized in 1N NaOH[b]	% [14]C in Precursor Recovered by TLC[c]
[2'-^{14}C]-Ferulic acid[d]	30.87 ± 6.48	2.14 ± 0.49
[2'-^{14}C]-Cinnamic acid[d]	19.33 ± 0.12	0.94 ± 0/07
L-[U-^{14}C]-Phenylalanine[e]	80.77 ± 11.44	0.72 ± 0.31

Source: Data provided by A. L. Pometto, III, and D. L. Crawford.
[a] Administered to Douglas fir through cut stems.
[b] 30 mg of ^{14}C-lignocellulose extracted with 10 ml of 1N NaOH for 48 h at room temperature (triplicate samples).
[c] NaOH extracts were acidified and extracted with diethylether. Ether extracts were then chromatographed on silica gel plates using butanol–ethanol–water (7:2:3). Non radioactive precursors were added prior to chromatography as carriers, and ^{14}C trapped in carrier spots was determined by liquid scintillation counting procedures.
[d] Xylem tissue.
[e] Phloem tissue.

using a nonamino acid lignin precursor (ferulate, coumarate, etc.) in the initial labeling step, or by removing contaminated proteins by protease treatments (Ellis et al., 1946; R. L. Crawford et al., 1979b; Reid, 1979). The former alternative is the more highly recommended.

Use of ^{14}C-ferulate or ^{14}C-coumarate has an added advantage in that their use allows preparation of specifically labeled ^{14}C-[LIGNIN]-lignocelluloses (R. L. Crawford et al., 1979b). For example, ^{14}C-ferualte may be synthesized specifically labeled in the side chain, aromatic ring, or methoxyl group (Kirk et al., 1975). Thus feeding of such specifically labeled lignin precursors to plants allows preparation of ^{14}C-[LIGNIN]-lignocelluloses labeled specifically in lignin's rings, side chains, or methoxyls (Haider and Trojanowski, 1975). Likewise, use of specifically labeled ^{14}C-coumarate allows preparation of ^{14}C-[LIGNIN]-lignocelluloses labeled specifically in the lignin ring or side chain carbons (Haider et al., 1977). Such specifically labeled ^{14}C-[LIGNIN]-lignocelluloses are very valuable for many studies of microbial lignin degradation (Chapters 4–6).

3.2.3   Preparation of ^{14}C-Labeled Milled-Wood Lignins (MWL's)

As discussed in Chapter 2, milled-wood lignins are among the best lignin preparations available for microbiological work. These lignins become particularly valuable when they are prepared so that they are ^{14}C-labeled. ^{14}C-MWL's may be prepared from ^{14}C-[LIGNIN]-lignocelluloses (D. L. Crawford, 1978) or from U-^{14}C-plant tissue. ^{14}C-MWL's are very promising substrates for microbiological studies. ^{14}C-MWL's are naturally synthesized polymers and may be prepared from many different plant species. Thus differences in biodegradability (^{14}C-MWL $\rightarrow$ ^{14}CO$_2$) between lignins of different plants may be examined without complicating interferences introduced by plant polysaccharides. Since plant lignins may be specifically labeled by feeding the plants specifically labeled precursors (e.g., ^{14}C-[RING]-, ^{14}C-[SIDE CHAIN]-, or ^{14}C-[METHOXYL]-ferulate), specifically tagged ^{14}C-MWL's may also be prepared. Thus ^{14}C-MWL's potentially offer many of the advantages of both ^{14}C-DHP's and ^{14}C-[LIGNIN]-lignocelluloses (R. L. Crawford et al., 1979b).

3.2.4   Preparation and Characterization of ^{14}C-Labeled Synthetic Lignins (DHP's)

As discussed in Chapter 2, the final step in the biosynthesis of lignin by vascular plants proceeds by a peroxidase-catalyzed free-radical polymerization of certain 4-hydroxycinnamyl alcohols. This polymerization is a complex phenol-coupling reaction, and can be reproduced in the laboratory using coniferyl alcohol, p-coumaryl alcohol, or sinapyl alcohol. In phosphate buffer containing H_2O_2, peroxidase, and vanillyl alcohol, a 4-hydroxycinnamyl alcohol polymerizes, forming a lignin-like dehydrogenative polymerizate (DHP) that precipitates from solution. Various authors have published procedures for preparing DHP's (Sarkanen and Ludwig, 1971; Kirk et al., 1975; Haider and Trojanowski, 1975; Haider et al., 1977), so a detailed protocol is not reproduced here. However, it is important to emphasize that the in vitro preparation of DHP's is a procedure requiring careful attention to detail (R. L. Crawford et al., 1979b), and experience with this type of controlled polymerization reaction is very helpful to those attempting a DHP synthesis (Crawford and Crawford, 1978).

DHP's containing a ^{14}C-label within the polymer can be prepared, and herein lies their major advantage to investigators studying lignin biodegradation. By using appropriately ^{14}C-labeled hydroxycinnamyl alcohols (usually ^{14}C-coniferyl alcohol or ^{14}C-coumaryl alcohol), Kirk et al. (1975) and Haider and Trojanowski (1975) have prepared ^{14}C-DHP's labeled specifically in the aromatic rings, propanoid side chains, or methoxyl groups. Thus by using the appropriate specifically labeled ^{14}C-DHP, the ability of microorganisms to attack particular portions of the lignin macromolecule may be examined (Kirk et al., 1975).

[14]C-Labeled synthetic lignins are the best models presently available for biodegradation research. Thorough spectroscopic and chemical analyses of DHP's by many investigators have shown that these model lignins contain essentially the same intermonomer linkages found in natural lignins; that is, DHP's and natural lignins are qualitatively very similar (Freudenberg and Neish, 1968; Sarkanen and Ludwig, 1971; Kirk et al., 1975). There are usually quantitative differences in linkage types when DHP's and natural lignins (milled-wood lignins) are compared (Sarkanen and Ludwig, 1971); however, considering the great structural variabilities of lignins throughout the plant kingdom, these differences between DHP's and natural lignin are probably not microbiologically significant. ^{14}C-DHP's are insoluble in water and have molecular weights ($\overline{M}_n$) of 1490–1600 (Kirk et al., 1975). Though chemically they are good lignin models, DHP's are unnatural in their complete lack of any associated carbohydrates. The significance of this lack of associated polysaccharides is discussed in later chapters, accompanying discussions of lignin degradation by particular microbial groups.

There are major disadvantages of using ^{14}C-labeled DHP's in biodegradation research, including (a) their preparation requires considerable experience and ability in the discipline of synthetic organic chemistry; (b) they are not associated with intact plant cell walls, as are natural lignins; and (c) they are expensive (in terms of costs of ^{14}C-labeled precursors of the required hydroxycinnamyl alcohols, and in labor costs).

3.2.5 ^{14}C-Labeled Lignins as Substrates for Biodegradation Studies

Methods employed to monitor the microbial degradation of ^{14}C-labeled lignocelluloses, DHP's, and MWL's are similar to procedures used by microbial ecologists to study biodegradation of organic compounds within environments such as soil or water (Crawford and Crawford, 1978; Haider and Martin, 1975; Hobbie and Crawford, 1969). Aliquots of soil or water, or a cell suspension prepared from a pure microbial culture, are incubated under defined conditions in the presence of a known quantity of ^{14}C-labeled substrate. Incubations typically are performed in closed vessels equipped with gassing ports that allow either continuous or intermittent culture aeration and gas-phase sampling (Crawford and Crawford, 1976; Kirk et al., 1975; Haider and Trojanowski, 1975). Evolved ^{14}CO$_2$ is trapped in aqueous NaOH or KOH, or in an organic base such as ethanolamine. Trapped ^{14}CO$_2$ is then quantified by liquid scintillation counting (LSC) techniques. Other radioactive gases such as ^{14}CH$_4$ may be quantified by trapping them separately (Ferry and Wolfe, 1976) from ^{14}CO$_2$, or by gas-chromatography/gas-flow proportional counting techniques (Hackett et al., 1977). ^{14}C-Labeled catabolic products released to the growth medium as soluble organic compounds may be quantified in total by LSC procedures (Crawford and Sutherland, 1979), or identified and/or isolated by analytical and preparative chromatography techniques (Keith et al., 1978).

3.3   Use and Relevance of Model Compounds for the Study of Lignin Biodegradation

It is an exceedingly difficult task to study the biochemical mechanisms whereby microorganisms degrade lignin. A primary reason for this difficulty is the pronounced structural complexity of the lignin molecule (Chapter 2). Unlike other biopolymers, lignin contains no readily hydrolyzable bond recurring at periodic intervals along a linear backbone. Instead, lignin is a three-dimensional, amorphous polymer containing many different stable carbon-carbon and ether linkages between phenylpropanoid monomeric units. Thus it is difficult to design experiments to ascertain what specific enzymic transformations are occurring during microbial decay of lignin. Usually, at best, gross chemical alterations (e.g., by quantification of various functional groups) in the lignin polymer before and after decay may be measured (Kirk and Chang, 1975).

Theoretically, one way to circumvent this problem of chemical complexity is to study the microbial degradation of simple lignin model compounds of known chemical structure. In this experimental approach, low-molecular-weight compounds that contain chemical structures known to occur in lignin are used (Figure 6.14). It is then assumed that what is learned concerning the biochemistry of degradation of these lignin models is relevant to catabolic mechanisms of lignin biodegradation (Muranaka et al., 1976).

It has not been established with certainty that the general use of lignin model compounds as substrates for the study of lignin biodegradation is a valid experimental approach. It is clear that microorganisms isolated for their ability to degrade lignin models often do not degrade the lignin macromolecule. However, known lignin degraders generally are able to decompose many lignin models. The major criticism of the use of model compounds to study lignin biodegradation is that the models are usually low-molecular-weight, water-soluble compounds, while lignin is a macromolecular, water-insoluble substance. Enzymes used by microorganisms to degrade models (Chapter 6) are typically intracellular and highly substrate specific, and are unlikely to attack an extracellular, complex polymer such as lignin. Enzymes that attack the lignin macromolecule thus may be very different (extracellular, nonspecific, etc.) from those that attack models (Haars and Hüttermann, 1980); however, both groups of enzymes may perform the same types of chemical transformations (demethylations, hydroxylations, ring fission, etc.).

Despite these criticisms it is likely that in certain instances the use of model compounds is a valid experimental approach to the study of lignin biodegradation. This is particularly true of recent work with the white-rot fungus *Phanerochaete chrysosporium*. This work and work concerning the biochemistry of lignin degradation are discussed in depth in Chapter 6.

Chapter **4**

LIGNIN-DEGRADING MICROORGANISMS

The true range of microbial groups that are able to degrade lignin has been the subject of much debate in recent years. Of the three generally recognized groups of saprophytic microorganisms [fungi, actinomycetes (filamentous bacteria), and eubacteria], only certain fungi previously had been thought to play a role in lignin degradation within natural environments (Kirk, 1971; Kirk *et al.*, 1977). However, it is now clear that lignin degradation is not an ability limited to the wood-rotting "white-rot" fungi. A variety of Ascomycetes and Fungi Imperfecti are now known to be lignin degraders, as are certain representatives of both the eubacteria and the actinomycetes.

4.1 LIGNIN DEGRADATION BY WHITE-ROT FUNGI

The white-rot fungi are usually thought of as those fungi that are able to extensively decompose all the important structural components of wood, including both cellulose and lignin (Kaarik, 1974). In addition, white-rot fungi are characteristically producers of extracellular phenoloxidases (enzymes which are probably involved in the processes of lignin degradation; Chapter 6), while nonligninolytic fungi usually are not (Kirk, 1971). There are several hundred species of white-rot fungi belonging to a variety of fungal families among the Holobasidiomycetidae, including the Agaricaceae, Hydnaceae, Corticiaceae, Polyporaceae, and Thelephoraceae (Kirk, 1971). There are also a few Ascomycetes in the Xylariaceae that cause a white-rot type of wood decay (Kirk, 1971). Of all the ligninolytic groups of microorganisms, the white-rot Basidiomycetes are probably the most efficient of all known lignin degraders. Under the proper environmental conditions, white-rot fungi completely degrade all structural components of lignin, with ultimate formation of CO_2 and H_2O (Cowling, 1961). A representative list of white-rot fungi is shown in Table 4.1.

Though all white-rot fungi are lignocellulose degraders, different species of white-rotters decay the various components of wood at different rates. For example, Polyporus berkeleyi removes the lignin from wood in preference to cellulose or hemicellulose (Kawase, 1962), as does Pycnoporus cinnabarinus (Ander and Eriksson, 1977). In fact, the latter fungus has been shown to degrade as much as 12.5% of the lignin in pinewood blocks (supplemented with malt extract) without loss of cellulose or mannan (Ander and Eriksson, 1977). Other white-rot fungi that under certain growth conditions degrade lignin preferentially to cellulose include Fomes ulmarius (=Rigidoporus ulmarius), Polyporus resinosus (Kirk, 1973; =Ischoderma resinosum), Pleurotus ostreatus, Phlebia radiata, and Merulius tremellosus (Ander and Eriksson, 1977). Ander and Eriksson (1977) have suggested that white-rot fungi that preferentially degrade lignin may be found characteristically among those fungi that produce large amounts of phenol oxidases. This hypothesis supports the earlier suggestion of Ishikawa et al. (1963) that phenoloxidase-rich white-rot fungi degrade more lignin than phenoloxidase-poor white-rot fungi. However, certain exceptionally good lignin-degrading white-rot fungi (e.g., Phanerochaete chrysosporium) produce barely detectable levels of phenoloxidases (Kirk and Fenn, 1979).

Many white-rot fungi deplete the lignin and carbohydrate components of wood at about the same proportional rates. Among these are Polyporus versicolor (Cowling, 1961; =Coriolus versicolor) and Ganoderma applanatum (Kirk, 1973). Another strain of Ganoderma applanatum (=Fomes applanatum), at the other extreme, has been shown to degrade the carbohydrates in wood more rapidly than the lignin (Kawase, 1962).

It is important to note that the experimental conditions used, particularly the composition of the growth medium employed, greatly influence the results of investigations of wood component degradation patterns. For example, Ander and Eriksson (1977) found that malt extract is generally stimulatory toward fungal lignin degradation as compared to other nitrogen sources. Also, Keyser et al. (1978) and Kirk et al. (1978) have shown that the concentration of nutrient nitrogen in the growth medium greatly influences DHP-lignin degradation rates by Phanerochaete chrysosporium (Chapter 6). The same authors also found that other culture parameters such as the O_2 concentration above cultures, the presence or absence of a readily utilizable growth substrate other than lignin, the medium pH, and the presence or absence of agitation had marked effects on lignin degradation rates of the Phanerochaete (Chapter 6). Even some common bufferring agents may inhibit fungal lignin degradation (Fenn and Kirk, 1979). Thus it is important

TABLE 4.1 Representative Species[a] of White-Rot Fungi[a]

Fungus	Reference
Phanerochaete chrysosporium (=*Sporotrichum pulverulentum*)	Kirk et al. (1975); Lundquist et al. (1977)
Phanerochaete velutina	Ander and Eriksson (1977)
Polyporus abietinus (=*Hirschioporus abietinus*)	Day et al. (1949)
Polyporus versicolor (=*Coriolus versicolor*)	Cowling (1961); Kawase (1962); Pelczar et al. (1950); Kirk and Highley (1973); Haider and Trojanowski (1975)
Polyporus dichrous (=*Gloeoporus dichrous*)	Ander and Eriksson (1977); Selin et al. (1975)
Polyporus adustus (=*Bjerkandera adusta*)	Sundman and Näse (1971)
Polyporus brumalis	Sundman and Näse (1971)
Polyporus picipes (=*P. badius*)	Sundman and Näse (1971)
Polyporus resinosus (=*Ishnoderma resinosus* and *P. benzoinus*)	Kirk and Moore (1971); Ander and Eriksson (1977)
Polyporus berkeleyi	Kirk and Moore (1971); Kawase (1962)
Polyporus giganteus (=*Polypilus giganteus*)	Kirk and Moore (1972)
Polyporus frondosus	Kirk and Moore (1972)
Poria subacida (=*Perenniporia subacida*)	Day et al. (1949)
Poria ambigua (=*P. latemarginata*)	Ander and Eriksson (1977)
Pleurotus ostreatus	Hiroi and Eriksson (1976); Sundman and Näse (1971); Kirk and Moore (1972); Haider and Trojanowski (1975)
Bjerkandera adusta	Ander and Eriksson (1977)
Pycnoporus cinnabarinus	Ander and Eriksson (1977)
Trametes zonata	Ander and Eriksson (1977)
Trametes hirsuta (=*Corrolus hirsutus*)	Ander and Eriksson (1977)
Trametes versicolor (=*Coriolus versicolor*)	Ander and Eriksson (1977); Sundman and Näse (1971); Selin et al. (1975)
Trametes pini (=*Phellinus pini*)	Sundman and Näse (1971)

40

Pholiota mutabilis	Ander and Eriksson (1977)
Pholiota spectabilis	Ander and Eriksson (1977)
(=*Gymnopilus spectabilis*)	
Pholiota squarrosa	Sundman and Näse (1971)
Cerrena unicolor	Ander and Eriksson (1977)
Phellinus isabellinus	Ander and Eriksson (1977)
Phlebia gigantea	Ander Eriksson (1977)
Phlebia radiata	Ander and Eriksson (1977)
Merulinus tremellosus	Ander and Eriksson (1977)
Lycoperdon pyriforme	Ander and Eriksson (1977)
Fomes applanatum	Kawase (1962)
(=*Ganoderma applanatum*)	
Fomes ulmarius	Ander and Eriksson (1977);
(=*Rigidoporus ulmarius*)	Kirk and Moore (1972)
Fomes annosus	Sundman and Näse (1971);
(=*Heterobasidion annosum*	Ishikawa *et al.* (1963)
or *Fomitopsis annosa*)	
Fomes fomentarius	Sundman and Näse (1971)
Fomes ignarius	Sundman and Näse (1971)
(=*Phellinus igniarius*)	
Fomes pini	Ishikawa *et al.* (1963)
(=*Phellinus pini*)	
Marasmius androsaceus	Sundman and Näse (1971)
Marasmius scorodonius	Sundman and Näse (1971)
Armillaria mellea	Sundman and Näse (1971)
(=*Armillariella mellea*)	
Hypholoma capnoides	Sundman and Näse (1971)
Polystictus abietinus	Sundman and Näse (1971)
(=*Hirschioporus abietinus*)	
Lenzites butulina	Sundman and Näse (1971)
Panus conchatus	Sundman and Näse (1971)
Stereum purpureum	Sundman and Näse (1971)
(=*Chondrostereum purpureum*)	
Xanthochorus obliquus	Sundman and Näse (1971)
Ganoderma applanatum	Kirk and Highley (1973);
(=*Elfvingia applanatum*)	Kirk and Moore (1972)
Peniophora sp.	Kirk and Highley (1973)
Cryptoderma yamanoi	Kirk and Moore (1972)
Radulum casearium	Thiverd and Lebreton (1969)

---

[a] No attempt is made to be comprehensive in this list of species or references.

that such experimental parameters be well controlled when wood component degradation patterns of different fungi are compared.

4.1.1 Chemistry of Lignin Degradation by White-Rot Fungi

Despite much effort during the past 20 years toward understanding the
biochemistry of wood decay, relatively little was known until very recently
concerning the chemical mechanisms whereby white-rot fungi (or any microbial group) decompose lignin. During the past 3–5 years much new
information about the chemistry of lignin degradation by white-rot fungi has
been obtained through analytical comparisons of nondecayed (sound) and
white-rotted wood lignins. Other valuable information has come from physiological studies of specific white-rot fungi, studies of fungal degradation of
lignin model compounds, and isolation of lignin decay fragments from
white-rotted wood. Discussions of physiological and model compound studies are saved for Chapter 6. The following section discusses the major
aspects of what is known concerning the chemistry of white-rotted lignins.

Lignins decayed by white-rot fungi contain less carbon, methoxyl, and
hydrogen than corresponding sound lignins (Hata, 1966; Ishikawa et al.,
1963a; Kirk and Lundquist, 1970; and Kirk and Chang, 1975). White-rotted
lignins contain more oxygen, more carbonyl groups, and more carboxyl
groups than sound lignins. The microbiologically introduced carbonyl
groups are largely conjugated to aromatic rings, as are about a third of the
newly introduced carboxyl groups. The remainder of the new carboxyls are
conjugated to α,β-unsaturated aliphatic structures (Kirk and Chang, 1975).
Degraded lignins can be up to 25% deficient in methoxyl groups, but there is
no significant accompanying increase in phenolic hydroxyl groups (Kirk and
Chang, 1975). Tables 4.2–4.4 summarize major aspects of what is known
concerning elemental and functional group compositions of sound and
white-rotted lignins.

White-rotted lignins are still polymeric. In fact, molecular weights of
decayed lignins are similar to those of sound lignins (Kirk and Chang, 1974).
Thus white-rot attack on the lignin macromolecule probably does not involve extensive depolymerization prior to further decay (although some
low-molecular-weight lignin degradation fragments are lost to the growth
medium during wood decay). These observations, together with analytical
data such as those shown in Tables 4.2–4.4, indicate that white-rot fungi
oxidatively degrade lignin by (a) demethylation and/or hydroxylation of
aromatic rings to produce diphenolic structures, followed by (b) rapid
oxygenolytic cleavage of the polyhydroxylated aromatic nuclei, yielding
aliphatic products (probably α,β-unsaturated carboxylic acids) that can be
further degraded by hydrolytic and aldolytic reactions. These reactions
probably proceed within the intact polymer and are likely to be catalyzed by

TABLE 4.2   Elemental Composition of Sound and White-Rotted Lignins

Lignin	% C	% H	% O	Reference
Pine Björkman, sound	63.70	6.29	30.01[a]	Ishikawa et al. (1963a)
Pine Björkman, decayed by Polyporus versicolor (=Coriolus versicolor)	61.41	6.11	32.48[a]	Ishikawa et al. (1963a)
Spruce Björkman, sound	62.31	5.88	31.31[a]	Hata (1966)
Spruce, decayed by Poria subacida (=Perenniporia subacida)	60.44	5.67	33.89[a]	Hata (1966)

Source: Kirk (1971).
[a] Determined by difference

extracellular or cell surface bound oxygenases (Kirk and Chang, 1975; Kirk et al., 1977). This possibility is discussed in more detail in Chapter 6; however, it should be noted that Kirk and Lundquist (1970) found that a milled-wood lignin prepared from sweetgum decayed by Polyporus versicolor did not differ significantly from a corresponding sound lignin according to a variety of chemical analyses. This observation led the authors to speculate that during decay of sweetgum by Polyporus versicolor (=Coriolus versicolor) only a small part of the lignin is attacked at any one time, and the

TABLE 4.3   Functional Group Contents of Sound and White-Rotted Lignins

Lignin	Functional Groups (moles/C9 unit)				
	Conjugated Carbonyl	Carboxyl Groups	Phenolic Hydroxyl	Aliphatic Hydroxyl	Methoxyl
Spruce, Björkman (sound)	0.07	0.1	0.24	0.92	0.92
Spruce, extractive (sound)	0.07	0.06	0.38	0.86	1.07
Spruce, decayed by Coriolus versicolor	0.17	0.55	0.11	—	0.74
Spruce, decayed by Polyporus anceps	0.16	0.58	0.10	0.77	0.72

Source: Kirk and Chang (1974; 1975).

TABLE 4.4 C_9-Unit Formulas of Sound and White-Rotted Lignins

Lignin	C_9 Formulas	MW of C_9 Unit
Spruce, Björkman (sound)	$C_9H_{8.66}O_{2.75}[OCH_3]_{0.92}$	189.2
Spruce, extractive (sound)	$C_9H_{8.79}O_{2.89}[OCH_3]_{0.96}$	192.8
Spruce, extractive (sound)	$C_9H_{8.85}O_{2.68}[OCH_3]_{1.07}$	192.9
Pine, Björkman (sound)	$C_9H_{8.86}O_{2.58}[OCH_3]_{0.94}$	187.3
Spruce, decayed by *Coriolus versicolor*	$C_9H_{7.26}O_{3.95}[OCH_3]_{0.74}$	199.4
Spruce, decayed by *Polyporus anceps*	$C_9H_{7.70}O_{3.80}[OCH_3]_{0.72}$	198.8
Spruce, decayed by *Poria subacida* (=*Perenniporia subacida*)	$C_9H_{8.61}O_{3.34}[OCH_3]_{0.77}$	193.9

Source: Kirk and Chang (1974); Hata (1966).

lignin that is attacked is degraded and utilized before decay proceeds to the rest of the lignin molecule. This hypothesis is consistent with the operation of an extracellular, surface-active lignin-degrading enzyme system. Also, Haider and Grabbe (1967) reported that certain white-rot fungi attack the aromatic rings of lignin faster than the side chains. These authors observed that ^{14}C-[ring]-DHP's were converted by their fungi to $^{14}CO_2$ faster than ^{14}C-[side chain]-DHP's. This observation led to the suggestion that cleavage of lignin's aromatic rings occurred while they were still bound in the polymer. Kirk *et al.* (1975) also examined the degradation of ^{14}C-DHP's by white-rot fungi. They found that both *Coriolus versicolor* and *Phanerochaete chrysosporium* converted specifically labeled ^{14}C-DHP's to $^{14}CO_2$ in the following order of decreasing rate: methoxyl-^{14}C > side chain-^{14}C > ring-^{14}C. Since these results differ from those of Haider and Grabbe (1967), the question of whether lignin's rings are degraded while still bound in the polymer remained unverified. However, work by Kirk and Chang (1975) with nonradioactive lignins seems to support this hypothesis. These authors found that extracted, degraded lignins did not contain lignin structures of molecular weight less than 600, and these high-molecular-weight degraded lignins were substantially dearomatized. The loss of aromaticity and retention of high molecular weight in white-rotted lignins suggests cleavage of aromatic rings while they are still bound within the lignin macromolecule.

In recent work which has been reported only in preliminary form, Chen et *al.* (1979) characterized by gas chromatography and mass spectrometry more than 30 aromatic-aliphatic lignin decay fragments released during decay of sprucewood by *Phanerochaete chrysosporium.* Assigned structures of these decay fragments (Figure 4.1) are again highly supportive of a lignin degradation pathway involving cleavage of rings within an intact polymer.

Krisnangkura and Gold (1979) characterized a guaiacyl lignin (coniferyl alcohol DHP) following decay by *Phanerochaete chrysosporium.* Characterization of the decayed lignin was accomplished by permanganate oxidation and methylation of the resulting aromatic acids, followed by quantitative gas chromatography of these methylated acids (Erickson *et al.*, 1973). When the decayed polymer was compared to undegraded DHP, it was found that condensed ring structures (e.g., biphenyl moieties and aryl-alkyl C-C bonded structures) were enriched in the decayed lignin. These results indicate that such structures are more resistant to attack by *Phanerochaete chrysosporium* than are other linkages in lignin.

It is known that lignins contain some peripheral phenylpropanoid units in which the propanoid side chains are not bonded to an adjacent unit of the polymer. Unit number 1 of Figure 2.3 and units 6' and 15 of Figure 2.4 are examples of such nonbonded side chains. There is good evidence that these phenylpropanoid end groups are oxidized by white-rot fungi, with conversion of the C_6-C_3 structures to C_6-C_1 moieties (Figure 4.2). Apparently these end groups are oxidized to the level of etherified aromatic acids. This conclusion is supported by the observation that nitrobenzene oxidation of white-rotted lignins yields less vanillin and more vanillic acid than oxidation of nondecayed lignins (Hata, 1966; Haider *et al.*, 1964; Ishikawa *et al.*, 1963b; Kirk and Chang, 1975; Higuchi *et al.*, 1955). Also, as discussed previously, white-rotted lignins contain increased quantities of aromatic

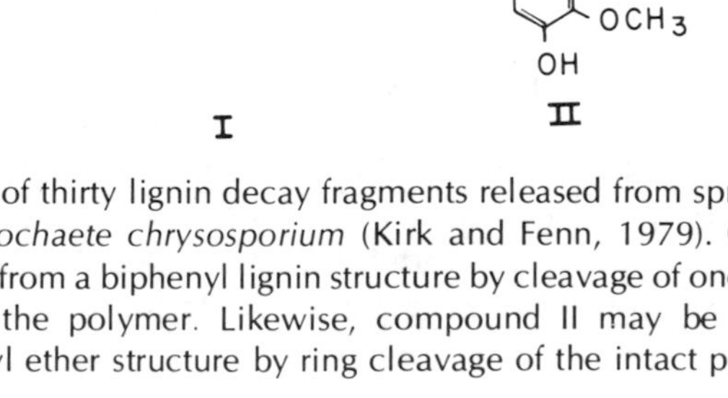

Figure 4.1 Two of thirty lignin decay fragments released from spruce wood during decay by *Phanerochaete chrysosporium* (Kirk and Fenn, 1979). Compound I was probably derived from a biphenyl lignin structure by cleavage of one ring while it was still attached to the polymer. Likewise, compound II may be derived from an arylglycerol-β-aryl ether structure by ring cleavage of the intact polymer.

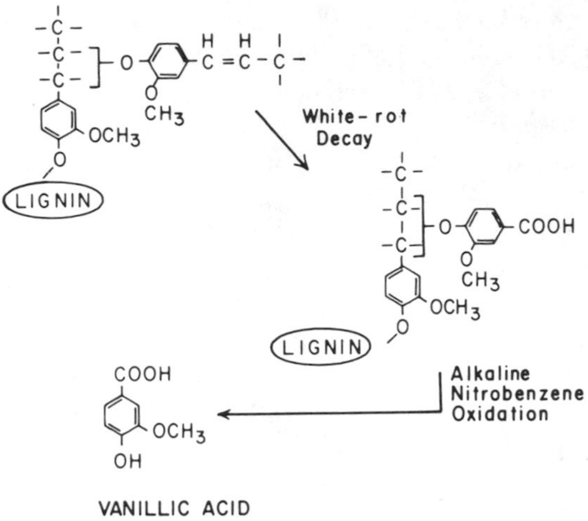

Figure 4.2   Degradation of phenylpropanoid end groups in lignin by white-rot fungi (Hata, 1966).

carboxyl groups as compared to sound lignins. As Kirk and Chang (1975) point out, the new vanillic acid moieties can arise by fungal oxidation of terminal phenylpropane units that occur originally in the natural polymer, or those resulting from fungal degradation of the macromolecule.

4.1.2   Cosubstrate Requirement for Lignin Degradation by White-Rot Fungi

Lignin decomposition by white-rot fungi is markedly affected by the presence or absence of carbohydrates in the growth medium. *Phanerochaete chrysosporium* and *Coriolus versicolor* do not degrade ^{14}C-DHP's to CO_2 in the absence of a readily utilizable growth substrate such as cellulose, glucose, glycerol, ethanol, or succinate (Kirk et al., 1976; Kirk and Fenn, 1979). Drew and Kadam (1979) have recently confirmed the obligate cosubstrate requirement for ^{14}C-kraft lignin degradation for three white-rot fungi including *Sporotrichum pulverulentum, Coriolus versicolor,* and *Phanerochaete chrysosporium* and for the imperfect fungus *Aspergillus fumigatus.*

Earlier work indicated that certain white-rot fungi actually utilize lignin as a sole source of carbon and energy (Gottlieb et al., 1950; Pelczar et al., 1950). These investigations involved the use of Brauns' native lignin prepared from spruce. Fungi examined included *Poria subacida* (=*Pereuniporia subacida*), *Polyporus versicolor* (=*Coriolus versicolor*), and *Polyporus abietinus* (=*Hirschioporus abietinus*). Somewhat later Ishikawa et al. (1963) reported that white-rotters in the genus *Fomes* could be adapted to growth on Brauns' native lignin as sole carbon source in minimal media. It is difficult to interpret these pieces of work because Brauns' lignin is probably not representative of the bulk of lignin in wood and may contain some low-molecular-weight components and/or carbohydrate contaminants that support artifactual fungal growth [though Gottlieb et al. (1950) "repurified" their lignin preparation by precipitating it from dioxane into ether 10 successive times]. It is instructive to note that Kirk et al. (1976) found that growth of *Phanerochaete chrysosporium* and *Coriolus versicolor* (=*Polyporus versicolor*) on milled-wood lignin (a more representative lignin preparation than Brauns' native lignin) was negligible.

Ander and Eriksson (1977) point out that lignin degradation has always been shown to occur simultaneously with degradation of at least one of the polysaccharides in wood. Their data show, however, that partially degraded lignins (e.g., kraft lignin) may be degraded by white-rot fungi to some extent in the absence of an additional energy source. Hiroi and Eriksson (1976) had made a similar observation somewhat earlier, but also noted a clearly demonstrable increase in kraft lignin degradation in the presence of cellulose. These authors observed very little degradation of milled-wood lignin by white-rot fungi in the absence of an alternative energy source.

Eriksson et al. (1980) studied the growth conditions in wood for three white-rot fungi and their cellulaseless mutants. They demonstrated that both water- and acetone-extractable substances in wood support the growth of cellulaseless mutants (permitting delignification in the absence of cellulose degradation). These and other observations supported the earlier observations that lignin in wood cannot be degraded by white-rot fungi unless a cosubstrate is simultaneously available.

In conclusion, it appears that lignins in their natural condition are not readily utilizable as a carbon/energy source by white-rot fungi. Evidence thus far strongly indicates that white-rotters probably do not degrade natural lignins significantly in the absence of an additional, more readily metabolizable carbon source (though some variances in specific growth substrate requirements between different white-rot species undoubtedly exist). As Kirk and Fenn (1979) point out, this apparent growth substrate requirement is unexpected, since lignin is highly reduced and potentially energy rich. These authors postulate that (a) for unknown reasons the net energy gain from lignin may be insufficient to support growth, or (b) lignin degradation may occur too slowly to sustain some minimal metabolic rate necessary for mycelial growth. They also point out that since lignin degradation by *Phanerochaete chrysosporium* and *Coriolus versicolor* only appears after nitrogen limitation of primary growth (Chapter 6), the lack of primary growth on lignin may simply reflect a cellular "control" event governing growth rather than the energy content of the lignin itself.

4.2 LIGNIN TRANSFORMATION BY BROWN-ROT FUNGI

Brown-rot fungi are usually defined as those wood-rotting fungi that decompose and remove wood carbohydrates, leaving a residue of modified lignin that is typically dark brown and almost equal in weight to the lignin in the original wood (Kirk, 1971; Ander and Eriksson, 1978). Brown-rotters are taxonomically very similar to white-rotters, being mostly Basidiomycetes. In fact, some fungal genera such as *Poria* contain representatives of both wood-rotting groups (Table 4.5).

The brown lignin residue left in brown-rotted wood is often called "enzymatically liberated" lignin (Schubert and Nord, 1950). Bray and Andrews (1924) were among the first to report that this brown-rotted lignin was chemically altered as compared to sound lignin. These authors reported a decreased methoxyl content in brown-rotted lignin. This observation has been confirmed by many investigators (Brown et al., 1968; Apenitis et al., 1965; Grohn and Deters, 1959; Pew and Weyna, 1962; Kirk and Adler, 1970; Kirk and Adler, 1969; Kirk, 1975).

Kirk and Adler (1969, 1970) examined brown-rotted lignins in relation to their decrease in methoxyl content and increase in phenolic hydroxyl con-

TABLE 4.5 Representative Species of Brown-Rot Fungi[a]

Species	Reference
Lenzites trabea	Kirk (1975)
(*=Gloephyllum trabeum*)	
Coniophora fusispora	Nilsson and Ginns (1979)
Coniophora prasinoides	Nilsson and Ginns (1979)
Coniophora suffocata	Nilsson and Ginns (1979)
Leucogyrophana arizonica	Nilsson and Ginns (1979)
Leucogyrophana mollusca	Nilsson and Ginns (1979)
Leucogyrophana olivascens	Nilsson and Ginns (1979)
Leucogyrophana pulverulenta	Nilsson and Ginns (1979)
Leucogyrophana romellii	Nilsson and Ginns (1979)
Merulius aureus	Nilsson and Ginns (1979)
Paxillus atrotomentosus	Nilsson and Ginns (1979)
Paxillus involutans	Nilsson and Ginns (1979)
Serpula incrassata	Nilsson and Ginns (1979)
Serpula lacrymans	Nilsson and Ginns (1979)
Poria vaillantii	Schubert and Nord (1950)
Lenzites sepiaria	Schubert and Nord (1950)
Poria monticola	Kirk and Highley (1973)
(*=Poria placenta*)	
Lentinus lepideus	Kirk and Highley (1973)

[a] Not a comprehensive listing.

tent. Their results were similar for lignins from spruce decayed by *Poria monticola* (=*Poria placenta*) and sweetgum decayed by *Lenzites trabea* (=*Gloeophyllum trabeum*). The brown-rotted lignins were extensively demethylated and contained o-diphenolic moieties. Both phenolic units (guaiacyl and syringyl) and nonphenolic (etherified) lignin units were demethylated. Some of the newly introduced phenolic hydroxyls in brown-rotted lignins are not generated by demethylation of methoxyls, but by direct hydroxylation of lignin rings *ortho* to propanoid side chains (Kirk et al., 1970; Kirk, 1975).

In addition to extensively demethylating lignins, brown-rot fungi also apparently introduce other chemical modifications into the lignin polymer. For example, brown-rotted lignins contain more carboxyl groups than corresponding sound lignins (Leopold, 1951; Enkvist et al., 1954; Kirk, 1975). The carboxyl groups in spruce lignin decayed by *Lenzites trabea* (=*Gloeophyllum trabeum*) were conjugated, about one third being aromatic and the remainder being α,β-unsaturated (Kirk, 1975). The same lignin contained about twice as much conjugated carbonyl (α to aromatic rings) as did sound lignin, and was somewhat lower in aliphatic hydroxyl content (Kirk, 1975). Table 4.6 compares the C_9 formulae of sound and brown-rotted spruce lignins.

Kirk (1975) concluded that brown-rot decay of spruce lignin by *Lenzites trabea* (=*Gloeophyllum trabeum*) was largely oxidative and that demethylation of both phenolic and nonphenolic units is the major degradative reaction. Demethylation reactions and ring hydroxylation reactions yield o-diphenolic structures; however, most of these new catechol moieties are further degraded by autooxidation or perhaps limited ring cleavage reactions. As with white-rot fungi, these observed oxidative changes were suggestive of the production of extracellular oxygenases by brown-rot fungi.

TABLE 4.6 C_9-Unit Formulas of Sound and Brown-Rotted Spruce Lignin

Lignin	C_9 Formula	MW of C_9 Unit
Milled-wood lignin, sound	$C_9H_{8.66}O_{2.75}[OCH_3]_{0.92}$	189.2
Extractive lignin, sound	$C_9H_{8.79}O_{2.89}[OCH_3]_{0.96}$	192.8
Brown-rotted lignin[a] (*Lenzites trabea* =*Gloeophyllum trabeum*)	$C_9H_{8.44}O_{3.75}[OCH_3]_{0.61}$	195.4

Source: Kirk (1975).
[a] The brown-rotted lignin was about 1 atom rich in oxygen, 0.3 atom poor in hydrogen, and 35% deficient in methoxyl content per C_9 "monomer" unit as compared to the corresponding sound lignins.

Thus brown-rot fungi appear initially to decay lignin in a manner similar to white-rot fungi (demethylations, hydroxylations, side chain oxidations, etc.); however, the brown-rotters apparently do not efficiently cleave lignin's hydroxyl-activated rings, or, if they do open the rings, they are unable to significantly decompose resulting aliphatic moieties (Kirk, 1971).

Brown-rot fungi have been examined for their abilities to convert ^{14}C-DHP's to $^{14}CO_2$. Kirk et al. (1975) examined degradation of ^{14}C-[side chain]-DHP, ^{14}C-[methoxyl]-DHP, and ^{14}C-[ring]-DHP by the brown-rotters Gloeophyllum trabeum (=Lenzites trabea) and Poria cocos. Conversion of these DHP's to $^{14}CO_2$ was, as expected, much less extensive than by white-rotters such as Phanerochaete chrysosporium and Coriolus versicolor. The brown-rotters slowly converted the DHP's to $^{14}CO_2$, with ^{14}C-[methoxyl]-DHP being degraded more rapidly than ^{14}C-[side chain]-DHP, which was in turn degraded more rapidly than ^{14}C-[ring]-DHP. As Kirk et al. (1975) point out, the low degradation of side chain- and ring-labeled lignins by the brown-rot fungi is fully in accord with previously discussed observations that brown-rot fungi do not substantially deplete lignin during wood decay. Their data are also consistent with the known propensity of brown-rot fungi to demethylate lignin.

It should be pointed out that, although conversions of ^{14}C-DHP's to $^{14}CO_2$ by Gloeophyllum trabeum and Poria cocos were much lower than by the two white-rot fungi, $^{14}CO_2$ recoveries were not insignificant (Kirk et al., 1975). After 600-h incubation under nonoptimized conditions Gloeophyllum trabeum converted 5% of the provided ^{14}C-[side chain]-DHP to $^{14}CO_2$ (Coriolus versicolor converted 22% of the side chain–labeled DHP to $^{14}CO_2$). Manipulation of culture conditions might increase the side chain degradation beyond 5%; thus it appears that brown-rotters may be able to degrade lignin more extensively than previously believed, though certainly not as extensively as white-rotters. Haider and Martin (1979) have also demonstrated quite a significant conversion of ^{14}C-lignins to $^{14}CO_2$ by the brown-rotter Gloeophyllum trabeum. As Kirk and Fenn (1979) point out, these observations raise the question of why lignin is not depleted significantly as wood is decayed by brown-rot fungi.

Our unpublished observations indicate that there is probably a significant depletion of lignin in spruce during decay by Gloeophyllum trabeum. As Figure 4.3 shows, Gloeophyllum trabeum converts specifically side chain-labeled ^{14}C-[LIGNIN]-spruce to $^{14}CO_2$ at a significant rate, although more slowly than the white-rotter Phanerochaete chrysosporium. It seems likely that much of the discrepancy between observations of lignin removal from wood by Gloeophyllum based on weight loss and/or Klason analyses versus lignin losses observed by monitoring $^{14}CO_2$ evolution from ^{14}C-lignins can be explained based on what is known concerning the oxidative fungal modifications of brown-rotted lignins. Though significant amounts of lignin

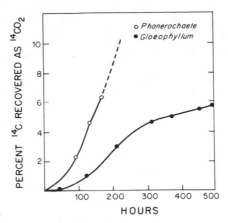

Figure 4.3 Degradation of ^{14}C-[LIGNIN]-spruce by *Gloeophyllum trabeum* and *Phanerochaete chrysosporium*. The *Phanerochaete* was grown in a defined medium containing 2 mM nitrogen (NH$_4$C1) and buffered (phthalate) at pH 4.5 (Kirk et al., 1978). The *Gloeophyllum* was grown in a medium optimized for growth (Kirk, personal communication) containing 0.2 g/l glutamic acid, 0.4 g/l K$_2$HPO$_4$, 0.1 g/l MgSO$_4$·7H$_2$O, 0.01 g/l thiamine, and trace metals at pH 4.4. ^{14}C-[LIGNIN]-spruce [487 dpm/mg, side-chain labeled and characterized as described by Robinson and Crawford (1978)] was supplied as growth substrate at 0.2 g/l. Incubation flasks (three replicates) containing 100 ml of media were sealed and equipped with ports for periodic flushing and trapping of evolved $^{14}CO_2$ as detailed by Kirk et al. (1975).

carbon (particularly side chain carbons and methoxyl carbons) may be lost as CO_2, this weight loss may be compensated for by extensive incorporation of molecular oxygen into the residual polymer (Figure 4.4). Structure I represents sound spruce lignin with a calculated elemental composition of C = 60%, O = 34%, H = 6%. It has a OCH$_3$ content of 15%. These values are within the normal range for sound spruce lignin (Kirk, 1975). Structure II represents brown-rotted spruce lignin, with the indicated structural changes based on chemical analyses of spruce lignin decayed by *Gloeophyllum trabeum* (Kirk, 1975). It should be realized the structural changes indicated are somewhat exaggerated quantitatively because it becomes unwieldy to expand the structure labeled "LIGNIN" to a realistic size; however, structure II has a calculated elemental composition of C = 54%, O = 41%, and H = 4.4%. It has a OCH$_3$ content of 10% (33% depleted as compared to structure I). These values are reasonable compared to those obtained for brown-rotted lignin (Kirk, 1975), particularly the decreases in H, C, and OCH$_3$ and increase in O with respect to structure I. In the transformation of I to II, CO_2 is produced from terminal side-chain and methoxyl carbons (potential sources of $^{14}CO_2$ from ^{14}C-lignins) with loss of 6.7% of the original carbon. Despite this carbon loss, the residual lignin (structure II) has actually

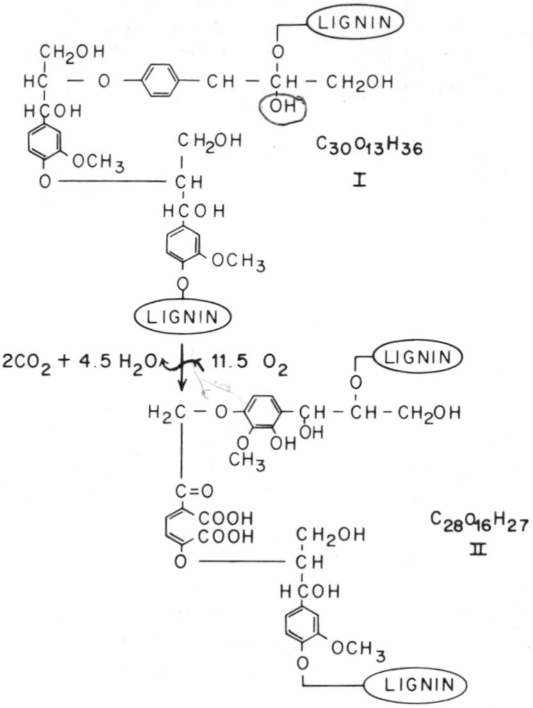

Figure 4.4 A hypothetical explanation for the conversion of lignin to CO_2 by *Gloeophyllum trabeum* without concomitant lignin weight loss.

gained about 2% in weight due to incorporation of oxygen atoms. Thus there is no apparent lignin loss by gravimetric procedures.

Kaplan and Hartenstein (1980) examined five brown-rot fungi for their abilities to convert ^{14}C-DHP lignins to $^{14}CO_2$. They found no significant $^{14}CO_2$ production (<0.5%) by any of their strains, including a strain of *Gloeophyllum trabeum*. Negative results were obtained with ^{14}C-[side chain]-, ^{14}C-[methoxyl]-, and ^{14}C-[ring]-labeled DHP's. The absence of some $^{14}CO_2$ evolution, particularly from ^{14}C-[methoxyl]-DHP, is surprising since it does not agree with the results of Kirk *et al.* (1975), Haider and Martin (1979), and our own experiments. Conflcting results in this case may be a result of differences in experimental (particularly cultural) conditions.

4.3   LIGNIN DEGRADATION BY SOFT-ROT AND OTHER FUNGI

4.3.1   The Soft-Rot Fungi

The soft-rot fungi attack moist wood, producing a characteristic softening of surfaces of the woody tissues (Savory, 1954; Corbett, 1965; Käärik, 1974;

Ander and Eriksson, 1978). A number of fungal genera contain representatives that cause a soft-rot type of wood decay (Table 4.7). Most of these genera are ascomycetes and various Fungi Imperfecti.

Soft-rot fungi are capable of causing extensive weight losses in wood (Duncan, 1960). One soft-rotter, *Chaetomium globosum*, was reported to cause 92% weight loss of beechwood (Savory and Pinion, 1958), implying very extensive lignin degradation. Levi *et al.* (1965) also reported extensive delignification of beechwood by *Chaetomium globosum*, finding lignin losses of up to 45%. Lignin decayed by *Chaetomium globosum* is apparently methoxyl deficient (Seifert, 1966).

Eslyn *et al.* (1975) examined the changes in chemical composition of poplar, alder, and pinewoods caused by six soft-rot fungi isolated from pulp chip storage piles. All the fungi were shown to deplete the lignin in these woods. Of the six soft-rotters examined, four removed polysaccharides faster than lignin. Two of the fungi (a *Paecilomyces* sp. and an *Allescheria* sp.) depleted lignin faster than cellulose and hemicellulose.

Da Costa and Bezemer (1979) recently examined techniques for the laboratory-scale production of soft-rotted wood. These authors recommended a procedure using wood blocks decayed on a garden compost.

TABLE 4.7   Representative Species of Soft-Rot Fungi

Fungus	Reference
Pestalotia multidea	DeCosta and Bezemer (1979)
Humicola sp.	DeCosta and Bezemer (1979)
Emericellopsis minima	DeCosta and Bezemer (1979)
Acremoniella sp.	DeCosta and Bezemer (1979)
Chaetomium globosum	Haider and Trojanowski (1975)
Graphium sp.	Eslyn *et al.* (1975)
Monodictys sp.	Eslyn *et al.* (1975)
Paecilomyces sp.	Eslyn *et al.* (1975)
Papulospora sp.	Eslyn *et al.* (1975)
Thielavia terrestris	Eslyn *et al.* (1975)
Allescheria sp.	Eslyn *et al.* (1975)
Xylaria sp.	Kirk (1972)
Phialophora sp.	Kirk (1972)
Alternaria sp.	Kirk (1972)
Cephalosporium sp.	Kirk (1972)
Pullularia sp.	Kirk (1972)
Cytosporella sp.	Kirk (1972)
Pestalozzia sp.	Kirk (1972)
Sporocybe sp.	Kirk (1972)
Preussia sp.	Haider and Trojanowski (1975)
Stachybotrys sp.	Haider and Trojanowski (1975)

They reported weight losses of up to 23.4% using *Chaetomium globosum* to decay the eucalypt spotted gum.

Though the above evidence strongly implicates an important role for soft-rotters as lignin degraders, inadequacies in methodologies still left room for doubt as to the true lignin-degrading abilities of these fungi. Haider and Trojanowski (1975) used ^{14}C-labeled lignins to conclusively prove that soft-rot fungi can cause a profound degradation of lignin. Soft-rotters including species of *Preussia, Chaetomium,* and *Stachybotrys* were able to convert ^{14}C-DHP's to ^{14}CO$_2$. All important structural elements of the DHP's (rings, side chains, and methoxyls) were converted to CO$_2$. The fungi also released ^{14}CO$_2$ from ^{14}C-labeled lignins of intact plant tissues (^{14}C-[LIGNIN]-maize).

There is essentially nothing known concerning the enzymatic mechanisms whereby soft-rot fungi convert lignin to CO$_2$.

4.3.2. Other Fungi

Several soil fungi that are not readily classifiable into specific decay groups are also known to be lignin decomposers. This group includes principally strains of *Aspergillus* and *Fusarium* and perhaps strains of *Endoconidiophora* and *Alternaria*.

Drew and Kadam (1979) and Hall *et al.* (1979) have reported that a strain of *Aspergillus fumigatus* converts ^{14}C-labeled kraft lignin to ^{14}CO$_2$ using soluble starch as a cosubstrate. These authors observed up to 50% conversion of ^{14}C-kraft lignin to ^{14}CO$_2$ in 16 days, a lignin degradation much faster than observed with white-rot fungi such as *Coriolus versicolor* under similar incubation conditions. It would be of interest to determine whether or not this *Aspergillus* is also able to degrade other ^{14}C-labeled lignin preparations such as ^{14}C-DHP's.

Higuchi (1979) isolated a number of *Fusarium* species that readily degraded dehydropolymers of coniferyl alcohol. These organisms preferentially degraded the low-molecular-weight fractions of the DHP's, but were also able to decompose some of the higher molecular weight DHP components. A similar observation was made by Iwahara (1979) with other DHP-degrading *Fusarium* strains. The authors did not utilize ^{14}C-labeled DHP's in their work, so information as to the structural specificity of attack on the DHP's (rings vs. side chains vs. methoxyls) by the fusaria was not obtained.

Chattopadhyay and Nandi (1977) reported that a strain of *Fusarium moniliforme* (var. *subglutinans*) was able to degrade the lignin in malformed mango inflorescence. Using the Klason lignin analysis procedure, these authors reported lignin losses of up to 26.5% after 30–45 days of incubation.

Ledingham and Adams (1942) reported that certain *Fusarium* and *Alternaria* species were capable of degrading 12–18% of the lignin component of a Ca-lignosulfonate preparation. In another paper (Adams and Ledingham,

1942) they also reported that a wood staining soil fungus *Endoconidiophora adiposa* was able to degrade 10% of the lignin fraction of sulfite waste liquor. In both these studies a β-naphthylamine precipitation method of questionable accuracy was used to estimate residual lignin, and adsorption of lignin to fungal mycelia was not considered as an alternative explanation for lignin disappearance.

Guylas (1967) reported that pure cultures of soil fungi (e.g., *Penicillium* and *Fusarium* strains) can degrade up to 20% of the lignin in wheat straw and about 11% of a lignin isolated from wheat straw. As Kirk (1971) points out, in this study residual lignin was determined by a sulfuric acid method, and in the case of work with the isolated lignin it is not clear whether lignin-mycelial adsorption was circumvented.

The work cited in the preceding paragraph indicates that these little-studied soil fungi are probably capable of extensive lignin degradation, though much of the very early work with these organisms is of questionable value (Kirk, 1971). Before an understanding of the true importance of these fungi as lignin decomposers can be realized, additional work is required, preferably using as substrates ^{14}C-labeled DHP's and ^{14}C-[LIGNIN]-lignocelluloses. The organisms isolated by Higuchi (1979) and Iwahara (1979) would be particularly valuable to examine in this regard.

Kaplan and Hartenstein (1980) examined six different Fungi Imperfecti, including a strain of *Aspergillus fumigatus*, for their abilities to degrade ^{14}C-DHP lignins (^{14}C-lignin $\rightarrow$ ^{14}CO$_2$). None of their cultures showed an ability to convert specifically labeled DHP's to CO$_2$ over an incubation period of 2 months.

4.4   BACTERIAL DEGRADATION OF LIGNIN

Numerous bacteria have been reported to decompose lignin, as summarized in Tables 4.8 and 4.9. As Kirk (1971) has pointed out, in many of the earlier papers listed in Table 4.8, weaknesses in experimental methods may have led to erroneous or at least questionable conclusions. Even some of the recent reports listed in Table 4.8 are equivocal because of problems with (a) nonrepresentative or chemically modified lignins being used as growth substrates, and (b) methods used to estimate residual lignins in culture media (e.g., A_{280} methods, cf. Section 3.1). Ander and Eriksson (1978) largely agreed with Kirk (1971), concluding that available evidence indicated that some bacteria may mediate a certain amount of lignin degradation, but their true abilities in this regard were unclear. However, the reviews by Kirk (1971) and Ander and Eriksson (1978) were written prior to publication of the papers listed in Table 4.9. These papers report the abilities of various

TABLE 4.8   Reports Implicating Bacteria as Lignin Degraders (Nonisotopic Methods)

Bacteria	Lignin Preparation	Reference
Pseudomonas	Brauns' lignin	Sorenson (1962)
Flavobacterium	Brauns' lignin	Sorenson (1962)
Xanthomonas	"Alkali" lignin	Jaschhof (1964)
Micrococcus	"Alkali" lignin	Jaschhof (1964)
Mixed Culture	Lignocellulose	Boruff and Buswell (1936)
Mxied Culture	Lignocellulose	Virtanen and Hukki (1946)
Arthrobacter	Brauns' lignin	Cartwright and Holdom (1973)
Pseudomonas *Bacillus* *Aeromonas* *Flavobacterium* *Cellulomonas* *Arthrobacter* *Xanthomonas*	Acidolysis lignin	Odier and Monties (1977)
Pseudomonas	Milled-wood lignin	Kawakami (1976)
Xanthomonas	Acidolysis lignin	Odier and Monties (1978a)
Acientobacter *Pseudomonas* *Xanthomonas*	Acidolysis lignin	Odier and Monties (1978a)
Pseudomonas	Lignin sulfonates	Kawakami et al. (1975a)
Pseudomonas	Kraft lignin	Kawakami et al. (1975b)
Pseudomonas *Chromobacterium*	Ligninsulfonate	Ban and Glanser-Soljan (1979)
Aeromonas	Kraft lignin	Deschamps et al. (1980)

pure bacterial cultures to convert ^{14}C-labeled lignins to $^{14}CO_2$, and conclusively prove that bacteria are capable of lignin degradation.

Trojanowski et al. (1977), Haider et al. (1978), and Gradziel et al. (1978) have shown that certain pure cultures of bacteria, particularly *Nocardia* species, are able to decompose lignin and to assimilate lignin degradation products as a carbon source. These bacteria could release $^{14}CO_2$ from ^{14}C-labeled methoxyl groups, side chains, or ring carbons of coniferyl alcohol DHP's and from specifically labeled ^{14}C-[LIGNIN]-maize. Experiments showed substantial releases of $^{14}CO_2$ (e.g., up to 15% of the ^{14}C in ^{14}C-[side chain]-maize lignin within 15 days) by the nocardias (Trojanowski et al., 1977). Since all the label release experiments were performed in the presence of other carbon sources such as yeast extract, glucose, or vanillic acid, it is unclear whether the bacteria degrade lignin in the absence of an alternative energy source. Though numerous *Nocardia* strains degraded ^{14}C-DHP lig-

TABLE 4.9 Reports Demonstrating Bacterial Degradation of ^{14}C-Labeled Lignins

Bacteria	Lignin Preparation	Reference
Nocardia	^{14}C-DHP, ^{14}C-[LIGNIN]-corn stalks	Trojanowski et al. (1977)
Bacillus	^{14}C-[LIGNIN]-spruce, ^{14}C-DHP	Robinson and Crawford (1978)
Streptomyces	^{14}C-[LIGNIN]-fir, ^{14}C-milled-wood lignin	D. L. Crawford (1978)
Nocardia, Pseudomonas	^{14}C-DHP	Haider et al. (1978)
Nocardia	^{14}C-DHP	Gradziel et al. (1978)
Streptomyces	^{14}C-[LIGNIN]-fir	Crawford and Sutherland (1979)
Streptomyces	^{14}C-[LIGNIN]-fir	Phelan et al. (1979)
Unidentified Bacterium	^{14}C-DHP	Kaplan and Hartenstein (1980)

nins quite efficiently, they varied much more in their abilities to degrade ^{14}C-[LIGNIN]-maize. Nocardia DSM 1069 converted ^{14}C-[methoxyl]-maize lignin to ^{14}CO$_2$ in about 13% yield in 15 days, while the corresponding 15-day yield from ^{14}C-[methoxyl]-DHP lignin was about 10%. Nocardia autotropica DSM 43089 converted ^{14}C-[methoxyl]-DHP lignin to ^{14}CO$_2$ in about 14% yield during 15 days, but only small amounts of ^{14}CO$_2$ were released from ^{14}C-[methoxyl]-maize lignin after 15 days (Haider et al., 1978). Apparently some DHP-degrading strains of Nocardia do not efficiently attack lignin that is associated with an organized cell wall (Haider et al., 1978).

Robinson and Crawford (1978) have shown that a Bacillus strain is able to convert ^{14}C-[side chain]-lignin of spruce to ^{14}CO$_2$. ^{14}CO$_2$ release from lignin's 2'-side chain carbons occurred at an initial rate equivalent to that observed with many lignin-degrading fungi (over 11% recovery of ^{14}C as ^{14}CO$_2$ during the first 15 days of incubation). The Bacillus also converted ^{14}C-[ring]-DHP to ^{14}CO$_2$, though at a much slower rate than ^{14}C-[LIGNIN]-spruce. The authors hypothesized that the initial release of ^{14}CO$_2$ from the side chain–labeled spruce lignins was due to degradation of ^{14}C that had been incorporated into peripheral units of the lignin which were more susceptible to attack than highly condensed lignins. They also observed, however, that continued evolution of ^{14}CO$_2$ past 35 days indicated some decomposition of more resistant lignin units.

Phelan et al. (1979) have recently examined the abilities of six lignocellulose-decomposing *Streptomyces* strains to degrade specifically ^{14}C-lignin–labeled Douglas fir lignocelluloses. In this work substrates included those in which only lignin side chain or ring carbons contained ^{14}C. Degradation of these components was followed over a 1008-h incubation period and comparisons were made with degradation patterns of similar substrates labeled simultaneously in both side chain and ring carbons. Results showed that aromatic ring components were cleaved and a substantial percentage of the labeled ring carbons were released as ^{14}CO$_2$. In contrast, side chain components were attacked to only a limited degree. Similar patterns were observed for all six strains, although the overall abilities of each strain to attack lignin varied considerably. Examination of rates of ^{14}CO$_2$ evolution from side chain or ring-labeled lignins as compared to rates of ^{14}CO$_2$ release from dual side chain/ring-labeled lignins indicated that these streptomycetes attack lignin's ring components preferentially. As Haider and Grabbe (1967) have suggested, this preferential release of ^{14}CO$_2$ could indicate ring opening within an intact lignin polymer. The preferential oxidation of lignin ring carbons by these *Streptomyces* contrasts markedly with the patterns of degradation by *Nocardia*.

A strain of *Streptomyces badius* has thus far been the most active lignin-decomposing actinomycete examined. In the work by Phelan et al. (1979) this strain released about 13% of a ^{14}C-lignin-labeled lignocellulose as ^{14}CO$_2$ in 1008 h of incubation at 37°C. An additional 16% of the total ^{14}C was recovered as water soluble ^{14}C (as compared to 2.8% for uninoculated controls). Growth of this culture on specifically side chain- and ring-labeled lignin preparations showed that almost all the total lignin degradation observed was due to attack on aromatic rings. This was in contrast to five other streptomycete strains that preferentially degraded ring components, but also significantly degraded side chain components. None of the strains used in this study were grown on specifically methoxyl-labeled lignins, although demethylation reactions would be expected to precede cleavage reactions.

D. L. Crawford (personal communication) and co-workers recently prepared ^{14}C-kraft lignin and ^{14}C-milled-wood lignin from ^{14}C-[LIGNIN]-Douglas fir (*Pseudotsuga menziesii*) lignocellulose. They then compared the abilities of a particular strain of *Streptomyces viridosporus* T7A to convert these three substrates to ^{14}CO$_2$. This appears to be the first instance where an investigator has compared the abilities of a bacterial strain to mineralize several different types of lignin preparations, all of which were derived from a single plant species. As Table 4.10 illustrates, the streptomycete mineralized both ^{14}C-milled-wood lignin and ^{14}C-[LIGNIN]-lignocellulose quite extensively (11–12% recovery of ^{14}C as ^{14}CO$_2$ after 28-days incubation). However, Douglas fir kraft lignin was very resistant to decomposition by the actinomycete (<3% recovery of ^{14}C as ^{14}CO$_2$ after 28 days). These observations again point out that the use of kraft lignins in microbiological work is not equiva-

TABLE 4.10   Production of $^{14}CO_2$ from ^{14}C-Labeled Douglas Fir Lignins by
Streptomyces viridosporus Strain T7A[a]

| Lignin | dpm/mg | % ^{14}C Recovered as $^{14}CO_2$ | |
		Inoculated	Noninoculated
Lignocellulose	1321	10.94 ± 1.94	0.61 ± 0.06
MWL[b]	1947	12.15 ± 1.71	1.24 ± 0.53
Kraft (KL)[b]	2302	2.42 ± 0.53	1.54 ± 0.09

[a] Cultures were grown in triplicate at 45°C for 28 days in a medium containing 0.3% acid-hydrolyzed casein [Bacto-Vitamin Free Casamino Acids; see D. L. Crawford (1978)].
[b] MWL = milled-wood lignin [Björkman (1956; 1957)]. Growth substrate was 10% ^{14}C-MWL or KL and 90% unlabeled lignocellulose.

lent to the use of lignins that have not been chemically modified (cf. Section 2.3).

A recent finding by Odier and Monties (1978a), that a *Xanthomonas* strain decomposed dioxane-lignin as a sole carbon and energy source, is very interesting in several respects. These authors observed 77% degradation of lignin in minimal medium after 15 days of growth. In the presence of glucose lignin degradation was suppressed (23% degradation observed). This work implies that certain bacteria may not require a cosubstrate for lignin degradation. In another study (Odier and Monties, 1977) the same authors found that 18 of 85 aerobic bacteria tested could use dioxane-lignin as a sole source of carbon and energy. Twenty-one others degraded the lignin in the presence of glucose. Unfortunately the abilities of these lignin-degrading bacteria to attack other types of lignin substrates (particularly ^{14}C-labeled lignins) was not systematically examined. Another major finding of Odier and Monties (1978) was that the *Xanthomonas* decomposed wheat dioxane-lignin anaerobically in the presence of nitrate and glucose. Anaerobic respiration as a mechanism for lignin decomposition has not been reported previously. In fact, Hackett *et al*. (1977) have reported that ^{14}C-dehydropolymers were not degraded at all in a variety of natural materials incubated under anaerobic conditions. In both papers cited above, it could have been low-molecular-weight fractions of the extractive dioxane-lignins that were metabolized by the bacteria. However, the degree of total decomposition reported was very high, and considerable chemical alterations of residual lignins were observed. This was indicative of extensive attack on polymeric (though chemically modified via acidolysis) lignin. Confirmation of these observations using ^{14}C-labeled lignins would be very valuable.

It has been established by Crawford and Sutherland (1979) and Sutherland *et al*. (1979) that an actinomycete, *Streptomyces flavovirens,* has the ability to attack and destroy the integrity of both lignified and nonlignified cell walls within the inner bark of Douglas fir (Figure 3.1). Although streptomycetes had previously been shown to colonize woody tissues (King and Eggins, 1977), this was the first report to show that an actinomycete could decompose the intact cell walls of woody plant tissue. Woody tissues decayed for 12 weeks by S. flavovirens were not significantly depleted in lignin (carbohydrates were depleted markedly), but the overall evidence (D. L. Crawford et al., 1979) indicated that this actinomycete invaded and destroyed the intact plant cell walls and structurally modified the lignin within them. This is the first evidence to actually implicate a bacterium with attack on lignin within intact woody tissue, and it provides evidence to indicate a wood decay role for actinomycetes.

In summary, it is now clear that bacteria are capable of extensive lignin degradation. Preliminary work indicates that pure cultures of bacteria are able to degrade all structural components of lignin, including methoxyl groups, aromatic rings, and side chains. It appears, however, that the ability to decompose lignin is much less widely distributed among bacteria than among fungi.

4.5 FUNGAL-BACTERIAL ASSOCIATIONS DURING LIGNOCELLULOSE DECOMPOSITION

It appears that utilization of fungal-bacterial associations may often speed the degradation of lignincellulosic materials as compared to degradation rates observed for fungi in pure culture. For example, Blanchette and Shaw (1978a) observed significant increases in wood decay (weight loss) during 5-month decay treatments by combining bacteria (Enterobacter sp.) and yeasts (Saccharomyces bailii var. bailii and Pichia pinus) with wood-rotting basidiomycetes such as Coriolus versicolor, Hirschioporus abietinus, and Poria placenta. In this mutualistic relationship the bacteria were thought to increase fungal growth and catabolism by supplying vitamins or other growth-promoting substances to the fungi. The bacteria in turn were able to utilize wood decay products released by fungal attack on the woody cell walls. This mixed-culture approach has been suggested as a management tool for rapid decay of forest residues left by logging operations (Blanchette and Shaw, 1978b).

Chapter 5

LIGNIN BIODEGRADATION WITHIN NATURAL ENVIRONMENTS

5.1 LIGNIN BIODEGRADATION IN SOIL AND WATER

Both ^{14}C-[LIGNIN]-lignocelluloses and ^{14}C-DHP lignins have been used to investigate rates of lignin degradation within soil and water. Hackett et al. (1977) examined ^{14}C-DHP biodegradation in a variety of natural materials using specifically labeled synthetic lignins. They found no biodegradation of ^{14}C-labeled lignin to labeled gaseous products under anaerobic conditions (e.g., in anaerobic marsh and lake sediments, anaerobic soil, and rumen fluid). Aerobic degradation of ^{14}C-DHP's was extensive and varied with respect to the type of natural materials examined (e.g., soils, sediments, and steer bedding), sampling site, soil type and horizon, and temperature (Hackett et al., 1977). Their greatest observed extent of ^{14}C-DHP degradation occurred in a soil from Yellowstone National Park (42% conversion to ^{14}CO$_2$ in 78 days). Extents of ^{14}C-DHP mineralization in soils were found to correlate with factors such as soil organic carbon, organic nitrogen, nitrate nitrogen, exchangeable calcium, and exchangeable potassium.

Martin and Haider (1977) and Haider et al. (1977) studied the biodegradation in soil of both ^{14}C-DHP's and ^{14}C-[LIGNIN]-corn stalks. Studies of ^{14}C-DHP degradation to ^{14}CO$_2$ in a Greenfield sandy loam indicated the DHP methoxyls were more susceptible to biodegradation by soil microbes than were side chain or ring carbons. γ-Side chain carbons (^{14}CH$_2$OH) were more susceptible to biodegradation than 2-^{14}C-[side chain] or ^{14}C-ring carbons. Conversions of ^{14}C-DHP's to ^{14}CO$_2$ in the Greenfield loam were substantial (Table 5.1). Corn stalk lignins were slightly more biodegradable (^{14}C-lignin $\rightarrow$ ^{14}CO$_2$) than the synthetic lignins, and (as with ^{14}C-DHP's) ^{14}CH$_2$OH side chain carbons were degraded faster than β-side chain or ring carbons (Table 5.1). Extraction of soil humic acids and fulvic acids, though

TABLE 5.1   Decomposition of ^{14}C-Lignins After 28 Weeks in a
Greenfield Sandy Loam

Substrate	% ^{14}C Recovered as ^{14}CO$_2$
^{14}CH$_2$OH-DHP	33
2-^{14}C-[side chain]-DHP	17–18
Ring-^{14}C-DHP	19–23
2-^{14}C-[side chain]-corn stalk lignin	28–29
^{14}CH$_2$OH-corn stalk lignin	37–38
1-^{14}C-[side chain]-corn stalk lignin	31
Ring-^{14}C-corn stalk lignin	27–28

Source: Haider et al. (1977).

problematical because of the NaOH solubility of residual undegraded DHP and partially degraded natural lignins, after incubation of soil in the presence of ^{14}C-lignins, indicated that carbon atoms of both ^{14}C-DHP's and ^{14}C-corn stalk lignins were incorporated into soil humus. The same humification patterns were observed for differentially labeled DHP's and corn stalk lignins, but humification of the natural lignins occurred to a lesser extent than for the synthetic lignins. In some admittedly speculative calculations, Martin et al. (1977) presented turnover times for lignin in the Greenfield sandy loam they had chosen for study. Turnover times for various lignin preparations ranged from 9 to 21 years.

Martin and Haider (1979) examined the degradation of ^{14}C-DHP lignins and ^{14}C-corn stalk lignins by the microflora of a fertile sandy loam. These authors reached two major conclusions concerning the conversion of ^{14}C-lignins to ^{14}CO$_2$ by soil microorganisms: (a) lignins are relatively resistant to complete mineralization, with the greatest conversion to CO$_2$ occurring during the earlier stages of decomposition; and (b) lignin mineralization rates are dependent on the presence of favorable soil conditions, but are not significantly influenced by the addition of readily available organic residues (e.g., orange leaves). The experiments reported by Martin and Haider (1979) are especially significant in that incubations were continued for a full year, the longest period yet reported for such experiments.

D. L. Crawford et al. (1977a,b) have examined degradation of ^{14}C-[LIGNIN]-lignocelluloses and ^{14}C-[GLUCAN]-lignocelluloses by the microflora of soil and water. Rates of lignin and cellulose degradation were estimated by monitoring ^{14}CO$_2$ evolution from incubation mixtures over incubation periods of up to 1000 h. Observed rates of lignin degradation were slow in all cases, and similar degradation patterns were observed in both soil and water (D. L. Crawford et al., 1977a,b). As also observed by Hackett et al. (1977), individual soil or water samples varied greatly in their rates and extents of lignin degradation. Figure 5.1 illustrates in graphic form

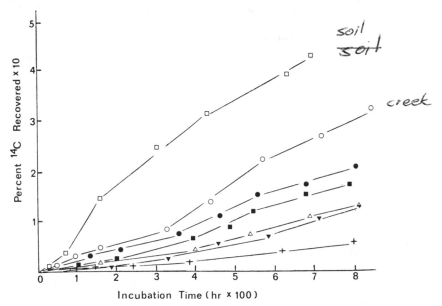

Figure 5.1　Rates of decomposition of ^{14}C-oak lignocellulose (^{14}C-lignin labeled) by selected soil and water microfloras [Crawford et al. (1977b). Dev. Ind. Microbiol. **19:** 44). Moist soil samples (2 g) or water samples (100 ml) in sterile ported flasks were incubated at room temperature with 10 mg (1.2×10^5 dpm) or 2 mg (2.4×10^4 dpm) of ^{14}C-oak lignocellulose, respectively (wood shavings with associated soil, □; GMU creek, ○; cornfield soil, ■; soil humus layer, △; white-rotted wood with soil, ▲; garden soil, +; Goose creek, ●).

the rates of microbial decomposition of ^{14}C-[LIGNIN]-oak by the microflora of selected soil and water samples.

　　Crawford et al. (1977a) compared soil-catalyzed degradation rates of the lignin components of six woods with degradation rates of their cellulose components. This was accomplished by monitoring evolution of $^{14}CO_2$ from identical soil samples supplemented with equal amounts of either ^{14}C-[LIGNIN]-lignocelluloses or the corresponding ^{14}C-[GLUCAN]-lignocelluloses. Lignin components of the six woods were shown to be decomposed by soil microfloras 4–10 times more slowly than were their corresponding cellulosic components. Figure 5.2 illustrates comparative degradation rates observed for the lignin and cellulose components of numerous woods, using soil taken from the floor of an oak–maple forest as a source of microbes (Crawford et al., 1977a).

　　To enrich and isolate microorganisms that are able to degrade resistant molecules such as lignin, it is desirable to identify those habitats that harbor the highest numbers of the desired microbes. Thus R. L. Crawford et al.

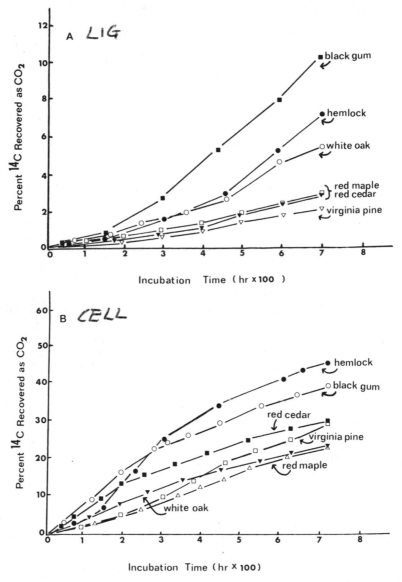

Figure 5.2 Decomposition of ^{14}C-labeled lignocelluloses by the microflora of an oak–maple forest soil (Crawford and Crawford, 1978). (A) ^{14}C-[LIGNIN]-lignocelluloses. (B) ^{14}C-[GLUCAN]-lignocelluloses.

(1977c) adapted the standard most probable number (MPN) procedure (Alexander, 1965) so that it may be used to enumerate lignin degraders in various natural or artificial environments. Evolution of $^{14}CO_2$ from dilution replicates containing ^{14}C-[LIGNIN]-lignocelluloses was used as a readily recognizable transformation for scoring replicates positive or negative for lignin degradation. Reference to standard MPN tables then permitted an estimation of absolute numbers of lignin degraders in the original soil or water used as an inoculum. Table 5.2 summarizes some typical numbers, estimated by the $^{14}CO_2$-MPN procedure, for lignin degraders in soil and water.

5.2 LIGNIN BIODEGRADATION WITHIN THE GUTS OF SOIL INVERTEBRATES

Many invertebrates ingest lignocelluloses as components of their diet. Among these are termites (e.g., *Reticulitermes flavipes*), isopods (e.g., *Oniscus asellus*), millipedes (e.g., *Oxidus gracilis*), snails (e.g., *Oxychilus draparnaldi*), slugs (e.g., *Deroceras reticulatum*), and earthworms (e.g., *Eisenia foetida*). Some interesting ecological and physiological questions may be asked concerning the fates of cellulose and lignin as they pass through the alimentary tract of such soil invertebrates. Though relatively little work has been published concerning biodegradation of lignocellulose in these highly specialized environments, available evidence indicates that cellulose may be degraded during passage of lignocellulose through an invertebrate's gut while lignin may be chemically modified but not significantly depleted. Jensen (1974) estimates that soil faunas actually metabolize less than 20% of the carbon in litter fall, with the remaining 80% being decomposed by the soil microflora. The mechanical disintegration of lignocellulosic plant materials is thought to be the major contribution of soil invertebrates to lignocellulose decomposition. Such disintegration enhances microbial attack on the more resistant plant polymers by improving aeration and water-holding capacities (Jensen, 1974; Lofty, 1974).

Seifert and Becker (1965) observed chemical changes of five species of

TABLE 5.2 Most Probable Numbers (MPN) of Lignin Degraders as Determined by $^{14}CO_2$ Evolution from ^{14}C-[LIGNIN]-lignocelluoses

Substrate	Inoculum	MPN[a]
Maple lignin	Lake water	7.9
Cattail lignin	Lake water	70.0
Maple lignin	Marsh water	49.0
Maple lignin	Forest soil	6.2×10^4
Cattail lignin	Marsh edge soil	$> 2.0 \times 10^6$

Source: Crawford and Crawford (1978).
[a] MPN per gram of dry soil or milliliter of water.

wood during passage through the alimentary canals of four species of termites. They reported lignin losses of 19–77%. However, as Esenther and Kirk (1974) point out, the high lignin digestion (77%) by the termite *Reticulitermes lucifugus* reported by Seifert and Becker (1965) was associated with an abnormal coprophagic situation and complete mortality of the termites. Thus their results are not representative of the true lignin-degrading ability of termites. Also, Butler and Buckerfield (1979) point out that the method used by Seifert and Becker (1965) to quantitate residual lignin (a 72% H_2SO_4 procedure) may be unreliable when applied to termite-modified lignin.

Esenther and Kirk (1974) examined the catabolism of aspen sapwood by the termite *Reticulitermes flavipes*. About 68% of the aspen wafers exposed to the termites was digested; however, there was essentially no change in the amount of sulfuric acid lignin after passage through the termites. Wood weight losses were explained by losses of glucan (91% loss), mannan (84% loss), and xylan (76% loss), rather than by losses of lignin. Even though not depleted or significantly demethylated, the lignin was chemically modified on passage through the termites. This was shown by acidolysis degradation of the residual lignin and analysis of acidolysis products by quantitative gas chromatography. Detailed analyses of chemical changes of the lignin were not performed.

French and Bland (1975) demonstrated incorporation of radioactivity into the body shells of two termites (*Nasutitermes exitiosus* and *Captotermes lacteus*) following feeding of [14]C-[LIGNIN]-eucalyptus (*Eucalyptus maculata*) wood. This result strongly implies that these insects (or their gut microfloras) are capable of lignin degradation. Butler and Buckerfield (1979) appear to have confirmed this observation. These authors found that *Nasutitermes exitiosus* extensively degraded both ^{14}C-DHP lignins (including methoxyl-labeled, side chain-labeled, and ring-labeled preparations) and [14]C-[LIGNIN]-maize. Release of ^{14}C as [14]CO_2 from the various DHP lignins ranged from 16.5 to 32.4% over incubation periods of 27–50 days. Release of ^{14}C as [14]CO_2 from maize lignins ranged from 63.1 (methoxyl-^{14}C) to 15.2% (ring-^{14}C) after 27 days of incubation. Controls were performed to confirm that lignin degradation occurred within the termites, rather than in voided feces. Butler and Buckerfield (1975) point out that their results raise questions concerning whether the usual assumption holds that termites' guts are anaerobic (Lee and Wood, 1971), since lignin degradation may be an obligately aerobic process (Hackett et al., 1977). Actually, the observations of Butler and Buckerfield (1979) point out that the question of whether lignin decomposition occurs in the absence of oxygen is still open.

A significant weakness in the paper by Butler and Buckerfield (1979), as well as in papers of other investigators (Martin and Haider, 1977; Haider et al., 1977; Neuhauser et al., 1978), is the lack of sufficient characterizations of their ^{14}C-DHP's, particularly as regards their molecular weights. Low-molecular-weight DHP's are more readily degradable than corresponding high-molecular-weight DHP's. Any particularly small DHP fragments (e.g., dimers and trimers) might be degraded by enzyme systems that are not actually components of a true lignolytic enzyme complex (Evans, 1977; Healy and Young, 1979). All authors should thus follow the example of Kirk et al. (1975) and report the approximate molecular weights of their DHP preparations.

Neuhauser et al. (1978) looked for conversion of specifically labeled ^{14}C-DHP lignins to ^{14}CO$_2$ during passage through the alimentary tracts of numerous soil invertebrates (isopods, snails, millipedes, slugs, and earthworms). Such a conversion would have been indicative of lignin degradation within the invertebrates' guts. The invertebrates were found unable to degrade ring-[^{14}C], methoxyl-[^{14}C], and side chain-[^{14}C]-lignin to ^{14}CO$_2$ over 10 days, thus providing strong evidence that soil invertebrates are not capable of extensive lignin degradation. Chemical transformations of lignin (short of CO$_2$ formation), however, would have gone unnoticed in these experiments, as the residual polymers were not critically analyzed. It is worthwhile to note that Neuhauser et al. (1974, 1976a, 1978) and Neuhauser and Hartenstein (1976a,b) have shown that isopods, earthworms, snails, slugs, and millipedes are able to convert simple aromatic compounds such as cinnamic acid 2-[^{14}C], vanillin 5-[^{14}C], benzoic acid ring-[^{14}C], and p-methoxyphenol methoxyl-[^{14}C] to ^{14}CO$_2$ and to incorporate ^{14}C from these types of compounds into their body tissues. Kaplan and Hartenstein (1978) and Neuhauser and Hartenstein (1976b) have provided some evidence to indicate that the actual cleavage of aromatic rings within invertebrates should be attributed to microbial enzymes, and not to invertebrate enzymes per se. The physiological basis for the lack of lignin degradation by invertebrates is as yet unexplained.

A wealth of information concerning the ecology of lignocellulose biodegradation within natural environments is found in the two volume set of papers Biology of Plant Litter Decomposition (C. H. Dickinson and G. J. F. Pugh, eds.), published by Academic Press (e.g., see Jensen, 1974). These volumes discuss much information not covered in the present discussion, and are highly recommended reading for readers who want to delve more deeply into this interesting area of ecology.

Chapter 6

BIOCHEMISTRY AND MICROBIAL PHYSIOLOGY OF LIGNIN BIODEGRADATION

6.1 MICROBIAL CATABOLISM OF BENZENOID MOLECULES: AN OVERVIEW

Lignin is a polymer of phenylpropanoid units; therefore its biological decomposition by necessity involves microbial attack on aromatic rings (cf. Section 4.1). Our understanding of the biochemical processes whereby lignin is decayed thus is contingent on our understanding of the mechanisms employed by microorganisms to dearomatize benzenoid nuclei. Though there is as yet only fragmentary evidence to support the contention, it is almost a certainty that mechanisms of microbial benzene ring degradation are similar for both lignaceous and nonlignaceous natural benzenoid substances. In other words, our understanding of the general mechanisms of benzene ring degradation by microorganisms should eventually lead us to an understanding of the mechanisms employed by microorganisms to degrade the benzene rings present in lignin.

6.1.1 Modern Concepts of Microbial Catabolism of Aromatic Molecules

It is now generally recognized that catabolism of benzenoid compounds is largely an aerobic process, with molecular oxygen being utilized in hydroxylation and ring-fission reactions, although information is also slowly accumulating concerning anaerobic catabolism of benzenoid molecules (Evans, 1977; Healy and Young, 1979).

The following is a brief outline of the most frequently investigated aerobic reaction sequences used by microorganisms (largely bacteria) to degrade aromatic molecules. This topic has been comprehensively reviewed by Chapman (1972), Dagley (1971, 1977) and Sugumaran and Vaidyanathan (1978). The reader is referred to these excellent reviews for a more detailed account of our present knowledge concerning aromatic catabolism by microorganisms. The present review is by necessity simply an overview of aromatic catabolism.

Catechol (o-dihydroxybenzene) and protocatechuic acid (3,4-dihydroxy-benzoic acid) have been the most commonly encountered ring-fission substrates used by microorganisms to dissimilate aromatic compounds. Generally, monosubstituted aromatic substrates such as phenol (hydroxy-benzene), benzoic acid, and mandelic acid (α-hydroxyphenylacetic acid) are catabolized via catechol, while para-disubstituted aromatic compounds such as p-hydroxybenzoic acid, p-hydroxymandelic acid, and p-cresol (1-hydroxy-4-methylbenzene) are degraded via protocatechuic acid. Table 6.1 lists some of the aromatic compounds dissimilated by way of catechol or protocatechuic acid.

Figure 6.1 illustrates the so-called "ortho-fission" (or "intradiol") pathways of catechol and protocatechuic acid dissimilation, which are collectively known as the β-ketoadipate pathway. The two catabolic sequences involve parallel chemistry, but different intermediate compounds, until they converge on a common metabolite, β-ketoadipate enol-lactone. This pathway is frequently encountered among bacteria of the genera Pseudomonas, Alcaligenes, Acinetobacter, Bacillus, Streptomyces, and Nocardia. The end products of this sequence are succinate and acetyl-CoA. The specificities of β-ketoadipate pathway enzymes are often high (Ornston, 1966a,b,c; Ornston and Stanier, 1966). The β-ketoadipate pathway has been reviewed by Stanier and Ornston (1973).

Figures 6.2 and 6.3 summarize alternative mechanisms for catabolism of catechol and protocatechuic acid. These pathways are known as "meta-fission" (or "extradiol") pathways. Ring-fission products of meta-fission oxygenases are semialdehyde derivatives and are bright yellow in neutral or basic solution. End products of the pathways illustrated are characteristically different from those of the ortho-fission pathways. Protocatechuic acid yields two pyruvate molecules while catechol yields one molecule each of pyruvate and acetaldehyde. These pathways are found among such bacterial genera as Bacillus, Pseudomonas, Arthrobacter, and Azotobacter. A novel version of meta-fission is found among certain species of Bacillus (see following). Enzymes of meta-fission pathways are often nonspecific, tolerating substitution of methyl or halogen substituents into the aromatic ring (Bayly et al., 1966; Tiedje et al., 1969). This is particularly true of enzymes of the catechol meta-fission pathway.

It has been shown that several strains of Bacillus circulans (R. L. Crawford, 1975a) and a strain of Bacillus macerans (R. L. Crawford et al., 1979)

TABLE 6.1　Catechol and Protocatechuate as Central Intermediates in Catabolism of Aromatic Compounds by Microorganisms

Compounds Metabolized Via			
Catechol		Protocatechuate	
Benzoate	(1)	m-Hydroxybenzoate	(12)
Phenol	(2)	Vanillate	(13)
Benzene	(3)	Isovanillate	(14)
o-Cresol	(4)	Veratrate	(15)
Mandelate	(5)	Vanillin	(16)
Salicylate	(6)	m-Cresol	(17)
Acetylsalicylate	(7)	p-Cresol	(18)
Naphthalene	(8)	p-Hydroxybenzoate	(19)
Phenanthrene	(9)	p-Aminobenzoate	(20)
Anthracene	(10)	Phthalate	(21)
Anthranilate	(11)	trans-Ferulate	(22)
Benzenesulfonate	(25)	α-Conidendrin	(23)
		p-Methoxybenzoate	(24)
		Cyclohexane carboxylate	(26)
		5-Hydroxyisophthalate	(27)
		Phenanthrene	(28)
		Phenylalanine	(29)

Sources: (1) Murray et al. (1972); (2) Johnson and Stanier (1971); (3) Gibson et al. (1968); (4) Ribbons (1966); (5) Kennedy and Fewson (1966); (6) Katagiri et al. (1965); (7) Grant (1971); (8) Evans et al. (1965); Yamamoto et al. (1965); (9) Davies and Evans (1964); Evans et al. (1965); (10) Davies and Evans (1964); (11) Higuchi and Sakamoto (1960); (12) Wheelis et al. (1967); (13–15) R. L. Crawford et al. (1973); (16) Toms and Wood (1970); (17) Stopher (1960) and Wheelis et al. (1967); (18) Bayley et al. (1966); (19) Henderson (1957); (20) Gibson (1968); (21) Ribbons and Evans (1960); (22) Henderson (1968); (23) Sundman (1962); (24) Buswell and Mahmood (1972); (25) Endo et al. (1977); (26) Blakley (1974); (27) Elmorsi and Hopper (1979); (28) Kiyohara and Nagao (1978); (29) Kishore et al. (1974).

degrade compounds such as benzoate, p-hydroxybenzoate, vanillate (4-hydroxy-3-methoxybenzoate), and p-hydroxyphenylpropionate via a novel pathway involving protocatechuate. Figure 6.4 illustrates this new and interesting sequence of reactions.

This novel pathway is initiated by a new *meta*-fission dioxygenase—protocatechuate 2,3-oxygenase. This enzyme cleaves the benzenoid nucleus of protocatechuate by inserting a molecule of oxygen between C_2 and

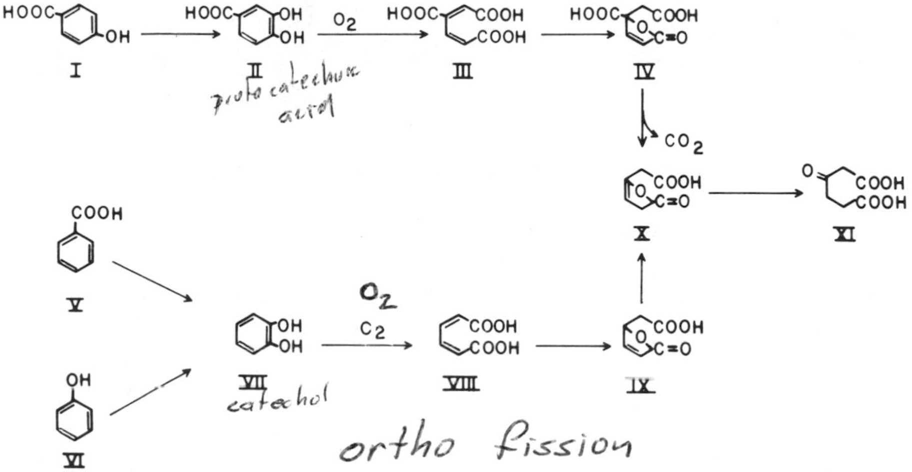

Figure 6.1 The β-ketoadipate pathway (Ornston, 1966a–c; Ornston and Stanier, 1966; Dagley, 1971).

C_3. All previously described protocatechuate oxygenases open the aromatic ring either between C_3 and C_4, or C_4 and C_5 (Dagley, 1971; Chapman, 1972). This observation therefore indicates an additional degree of versatility among soil microorganisms in degrading a classical metabolite of aromatic catabolism, protocatechuic acid.

Following protocatechuate 2,3-oxygenase, this novel pathway is similar to the *meta*-fission route of catechol degradation (Sala-Trepat et al., 1972),

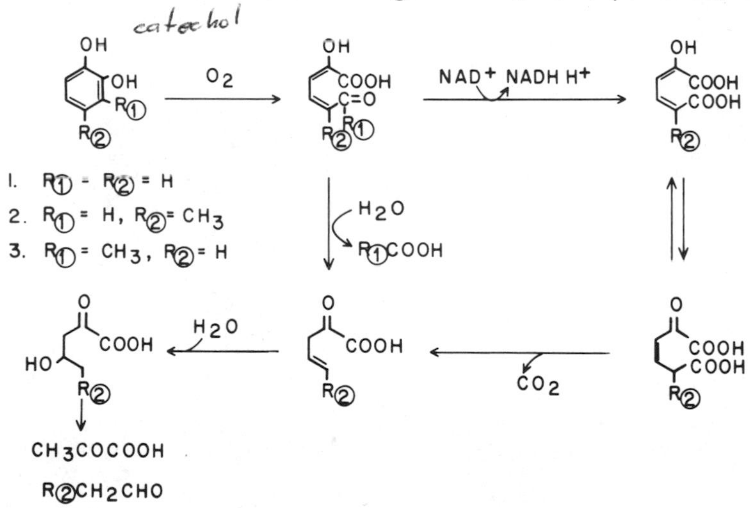

Figure 6.2 Catechol and methylcatechols meta-fission pathways (Bayly et al., 1966; Sala-Trepat et al., 1972; Sugumaran and Vaidyanathan, 1978).

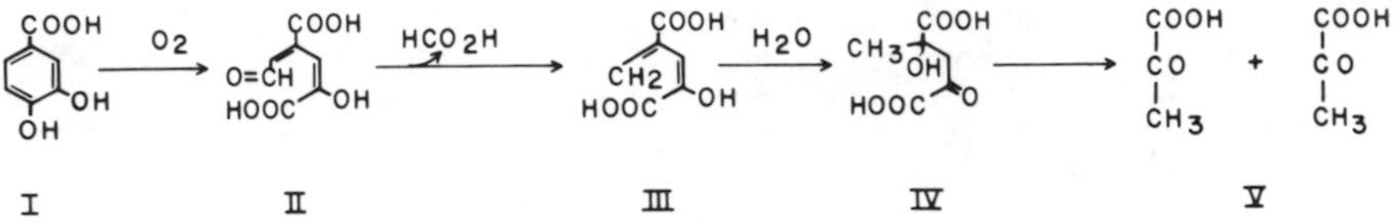

Figure 6.3 Protocatechuate meta-(4,5)-fission pathway (Dagley, 1971).

yielding one pyruvate and one acetaldehyde molecule per molecule of protocatechuate. Both a hydrolytic branch and an oxidative branch for the degradation of α-hydroxymuconic semialdehyde are present in cells; however, the oxidative branch (NAD+-dependent) is the predominant path, as it is in phenol-grown *Azotobacter* (Sala-Trepat et al., 1971).

We have recently purified a PCA-2,3-dioxygenase from a strain of *Bacillus macerans* following its growth on 4-hydroxybenzoate as sole source of carbon and energy (R. L. Crawford et al., 1979). Purification of the dioxygenase to electrophoretic homogeneity was accomplished by affinity chromatography using Sepharose beads with PCA covalently attached to a spacer molecule via the PCA carboxyl group. PCA-2,3-dioxygenase adsorbed to the affinity resin while other proteins did not. The dioxygenase was deadsorbed from the resin using a NaCl gradient.

Purification of the PCA-2,3-dioxygenase by affinity chromatography allowed firm proof that the conversion of PCA to 2-hydroxymuconic semialdehyde involves two enzymatic steps: ring-fission between C_2 and C_3, followed by rapid, enzymatic decarboxylation of 5-carboxy-2-hydroxymuconic semialdehyde (Figure 6.4).

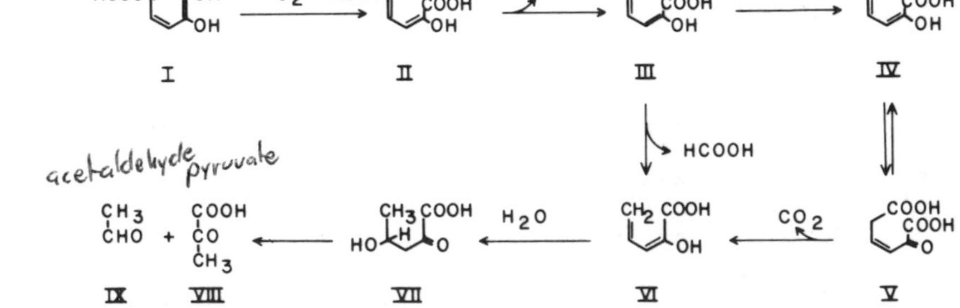

Figure 6.4 Protocatechuate 2,3-dioxygenase pathway of *Bacillus* species (R. L. Crawford, 1975a).

Homoprotocatechuic acid (3,4-dihydroxyphenylacetic acid) is a less frequently encountered ring-fission substrate. Figure 6.5 illustrates the chemistry of the homoprotocatechuate pathway. The chemistry of this pathway parallels that of the meta-fission pathway for catechol degradation (Figure 6.2). Aromatic compounds degraded via homoprotocatechuate include p-hydroxyphenylacetic acid, tyrosine, and perhaps phenylacetic acid. The homoprotocatechuate pathway has been observed among species of *Bacillus, Arthrobacter, Pseudomonas, Acinetobacter,* and *Alcaligenes.* Its end products are pyruvate and succinate (Sparnins and Chapman, 1976; Sparnins et al., 1974).

Gentisic acid (2,5-dihydroxybenzoic acid) is another frequently encountered substrate for ring-fission dioxygenases. Figure 6.6 illustrates the sequences of reactions known as the gentisate pathway(s). Aromatic compounds dissimilated via these series of reactions include m-hydroxybenzoic acid, p-hydroxybenzoate (Figure 6.13), p-hydroxyphenylpropionate, m-cresol (1-hydroxy-3-methylbenzene), and 2-naphthol (2-hydroxynaphthalene). Members of the bacterial genera *Vibrio, Pseudomonas, Bacillus, Micrococcus,* and *Moraxella* as well as plants and mammalian systems are known to

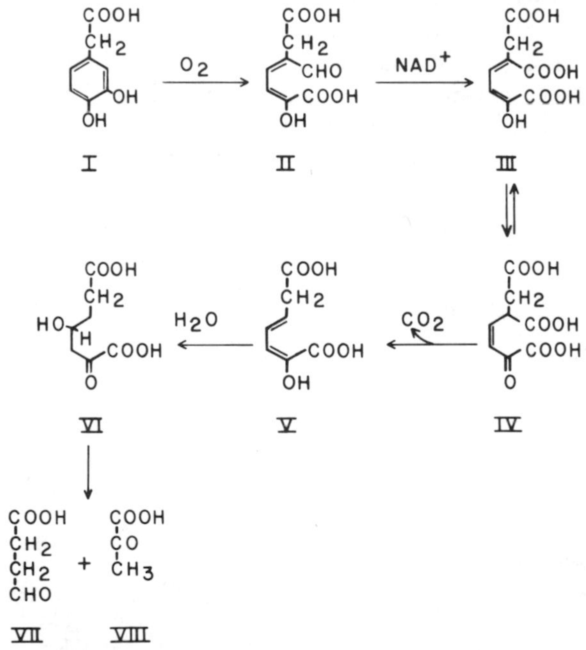

Figure 6.5 The homoprotocatechuate pathway (Sparnins et al., 1974; Sugumaran and Vaidyanathan, 1978). Used by permission of the American Society for Microbiology.

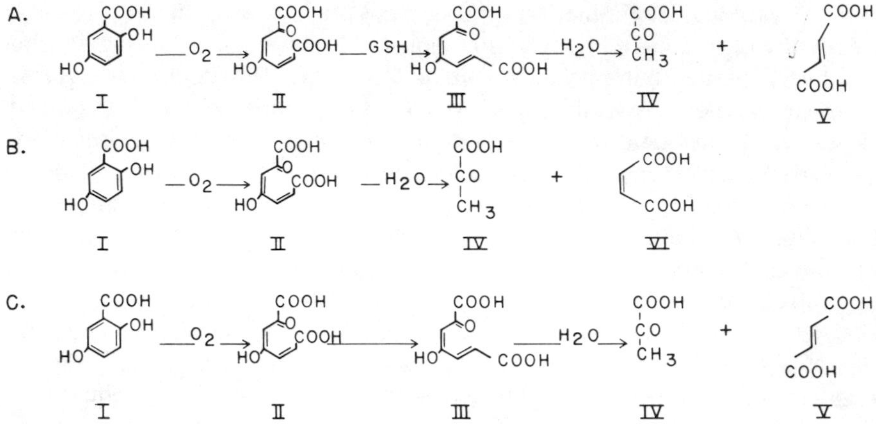

Figure 6.6 Microbial catabolism of gentisate (Crawford and Frick, 1977).

use these pathways. Gentisate pathway end products are fumarate (or sometimes maleate or D-malate) and pyruvate (Crawford and Frick, 1977; Lack, 1959; R. L. Crawford, 1975b).

Homogentisic acid (2,5-dihydroxyphenylacetic acid) is a somewhat less frequently encountered ring-fission substrate. Figure 6.7 illustrates the reactions of the homogentisate pathways. Aromatic compounds degraded via these routes include p-hydroxyphenylacetic acid and tyrosine. Homogentisate pathways have been observed among species of *Aspergillus, Bacillus, Pseudomonas,* and *Moraxella.* Their characteristic end products are fuma-

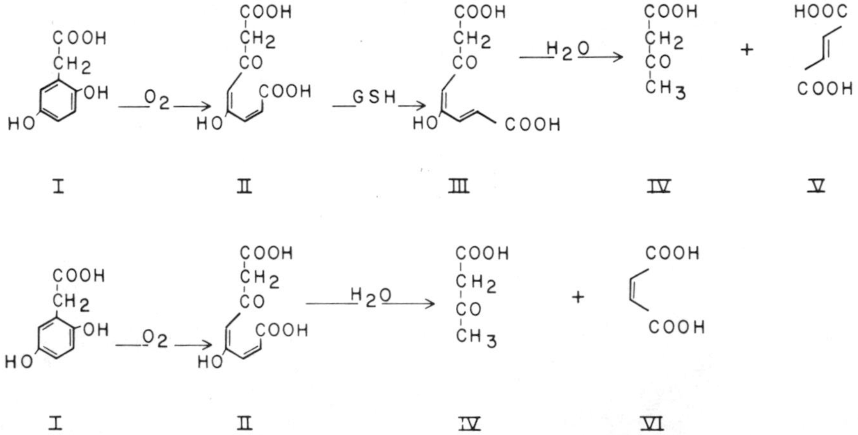

Figure 6.7 Microbial catabolism of homogentisate (Chapman and Dagley, 1962; R. L. Crawford, 1976b). Reproduced by permission of the National Research Council of Canada from the Canadian Journal of Microbiology, Volume 22, pp. 276–280 1976.

rate (or sometimes maleate) and acetoacetate (R. L. Crawford, 1976b; Chapman and Dagley, 1962).

Figure 6.8 illustrates some bacterial dioxygenases that are known to cleave dihydric phenols. These dioxygenases, in combination with various hydroxylases, give microorganisms catabolic capability and versatility not generally possessed by other groups of organisms.

Thus several generalizations concerning aromatic catabolism by microorganisms may be made. Aromatic compounds must first be converted to ring-fission substrates, which generally contain a minimum of two hydroxyl substituents oriented in an *ortho* or *para* relationship to one another. These ring-fission substrates are then cleaved by a class of enzymes known as dioxygenases to give aliphatic compounds that are further degraded to molecules that are readily funneled into the energy-yielding TCA cycle. Catabolism by these pathways of ring-substituted aromatic compounds, particularly those substituted with halogen atoms, generally requires dissimilatory pathways with enzymes of low specificity. Enzymes of *meta*-fission pathways are frequently more tolerant of such ring substitution than are enzymes of *ortho*-fission pathways. Exceptions to these general rules are known (Chapman, 1972; and the following).

6.1.2   Recently Discovered Novel Catabolic Reactions for Microbial Degradation of Benzenoid Molecules

Cleavage of Aromatic Ethers

Cleavage of aromatic ethers (e.g., methoxyl groups of compounds like vanillic acid; Ribbons, 1970) is usually accomplished by bacteria through the action of mixed-function oxygenases (or monooxygenases) that hydroxylate the first carbon of the etheric side chain, forming an unstable hemiacetal that decomposes to form a free phenol and an aliphatic aldehyde (Hareland et al., 1975; Ribbons, 1970). Alternatively, ether cleavage may result from hydroxylation of the aromatic ring carbon that bears the ether substituent, forming an unstable hemiketal that decomposes, yielding a free quinone and an aliphatic alcohol (Hareland et al., 1975). The only previously reported instance of the latter ether cleavage mechanism in bacteria involved hydroxylation of 4-hydroxyphenoxyacetate by a 4-hydroxyphenylacetate 1-hydroxylase, forming benzoquinone and glycolate (Hareland et al., 1975). In that instance 4-hydroxyphenoxyacetate was not the normal substrate for the 1-hydroxylase, but was a substrate analog. The *Pseudomonas* used did not grow on 4-hydroxyphenoxyacetate. Thus the ring-hydroxylation mechanisms for fission of aromatic ethers remained to be established as a true, catabolic pathway reaction.

We recently reported isolation of a *Bacillus* (strain PHPXAa-B) that grows

Dioxygenase reactions

on a 4-hydroxyphenoxyacetate as a sole source of carbon and energy (R. L. Crawford, 1978). The *Bacillus* initiates catabolism of 4-hydroxyphenoxyacetate by hydroxylation of C_1 of the aromatic nucleus, as shown in Figure 6.9. Thus we have established that bacteria may indeed cleave aromatic ethers by a mechanism that probably involves an intermediate hemiketal formed by hydroxylation of the ring carbon bearing the ether substituent.

Vanillate Decarboxylation by Soil Bacteria

We have in our culture collection five strains of *Bacillus megaterium* and one strain of a *Streptomyces* sp. that are able to convert vanillate to guaiacol in approximately quantitative yield (Figure 6.10). For example, when *Bacillus megaterium* D2d is suspended in buffer containing 0.01M vanillate, vanillate begins to disappear and guaiacol begins to be formed after a lag phase of about 2 h (required for induction of appropriate enzymes). After about 8 h the conversion of vanillate to guaiacol is quantitative. Inclusion of 1 mg/ml chloramphenicol in the buffer completely inhibits both the disappearance of vanillate and the formation of guaiacol.

Guaiacol is not usually considered a natural product, but is thought to be an industrial byproduct of industries such as pulp and paper manufacturing. Vanillate is easily isolated from natural environments such as decomposing wood. Thus our observation of microbial decarboxylation of vanillate to guaiacol shows that guaiacol is actually a natural compound.

We have modified the basic most probable number (MPN) procedure to ascertain the quantitative importance of *Bacillus* species as decarboxylating agents for vanillate. A medium was used that contained vanillate and yeast

---

Figure 6.8 Representative dioxygenases: (1) catechol 1,2-dioxygenase (Ornston, 1966a–c); (2) protocatechuate 3,4-dioxygenase (Ornston, 1966a–c); (3) catechol 2,3-dioxygenase (Bayly et al., 1966); (4) protocatechuate 4,5-dioxygenase (Dagley, 1971); (5) 2,3-dihydroxyphenylpropionate 1,2-dioxygenase (Dagley et al., 1963); (6) 2,3-dihydroxybenzoate 3,4-dioxygenase (Ribbons and Senior, 1970); (7) homoprotocatechuate 2,3-dioxygenase (Sparnins et al., 1974); (8) gentisate 1,2-dioxygenase (Crawford and Frick, 1977); (9) homogentisate 1,2-dioxygenase (Chapman and Dagley, 1962); (10) protocatechuate 2,3-dioxygenase (R. L. Crawford, 1975a); (11) 2,3-dihydroxy-*p*-cumate 3, 4-dioxygenase (DeFrank and Ribbons, 1977); (12) 2,3,5-trihydroxybenzoate 1,2-dioxygenase (Crawford and Olson, 1978); (13) 2,5-dihydroxy-4-methoxybenzoate 1,2-dioxygenase (Chapman, 1977); (14) gallate 4,5-dioxygenase (Sparnins and Dagley, 1975); (15) 2,5-dihydroxybenzaldehyde 4,5-dioxygenase (Scot and Beadling, 1974); (16) 3,4-dihydroxycinnamate 3,4-dioxygenase (Seidman et al., 1969); (17) homoprotocatechuate 3,4-dioxygenase (Subba Rao et al., 1971); (18) hydroquinone dioxygenase (Larway and Evans, 1965); for a recent compendum of dioxygenase catalyzed reactions, see Sugumaran and Vaidyanathan, 1978.

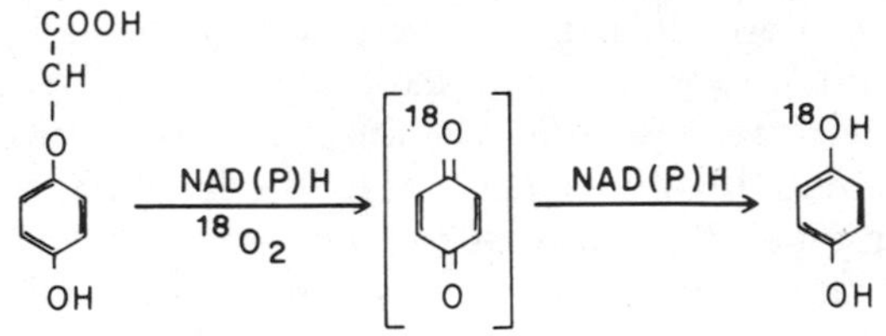

Figure 6.9 Hydroxylation of 4-hydroxyphenoxyacetate by a *Bacillus* (R. L. Crawford, 1978). For a proposed mechanism for this reaction, see Hareland *et al.* (1975).

extract as growth substrates. This medium was inoculated with pasteurized (80°C/10 min) dilutions of soil and natural waters (10-fold dilutions with five replicates per dilution) and incubated at 30°C for 3–4 weeks. At the end of the incubation period dilution replicates were acidified (pH 2) and extracted with ether. Ether extracts were examined by thin-layer chromatography for the presence of guaiacol. Replicates were scored positive if guaiacol was observed. Positive tubes provided an MPN for vanillate decarboxylators using standard statistical tables. Soil samples typically contained about 10^1–10^4 *Bacillus* spores per gram dry soil that were able to germinate and decarboxylate vanillate forming guaiacol. Thus decarboxylation of vanillate to guaiacol appears to be an important ability of soil bacilli (Crawford and Perkins-Olson, 1978b).

Catabolism of 3,5-Dihydroxybenzoate by a Bacillus

Numerous phenols are known to be substrates for aromatic-ring-fission dioxygenases, with protocatechuate and catechol being the most studied among the group (Dagley, 1971). The compound 2,3,5-trihydroxybenzoate is a potential ring-fission substrate for 3,5- or 2,3,5-substituted benzoates; however, until recently there were no reports indicating the existence of a specific 2,3,5-trihydroxybenzoate dioxygenase among microorganisms. We have discovered a novel catabolic pathway used by a strain of *Bacillus brevis* to degrade 3,5-dihydroxybenzoate. This new pathway involves dioxygenase-catalyzed fission of 2,3,5-trihydroxybenzoate between C_1 and C_2 (Figure 6.11; Crawford and Perkins-Olson, 1978a).

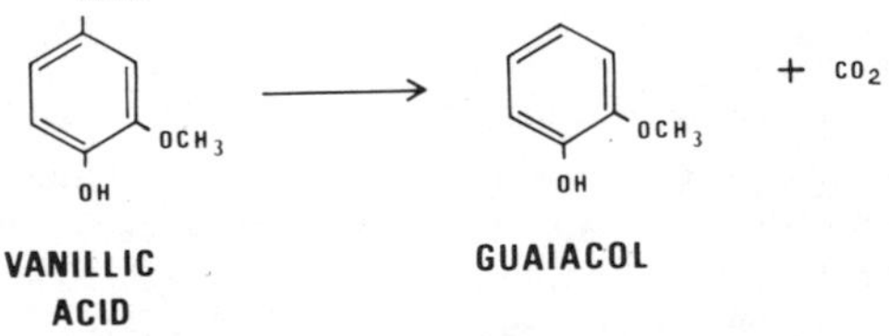

Figure 6.10 Bacterial decarboxylation of vanillate to form guaiacol (Crawford and Olson, 1978b).

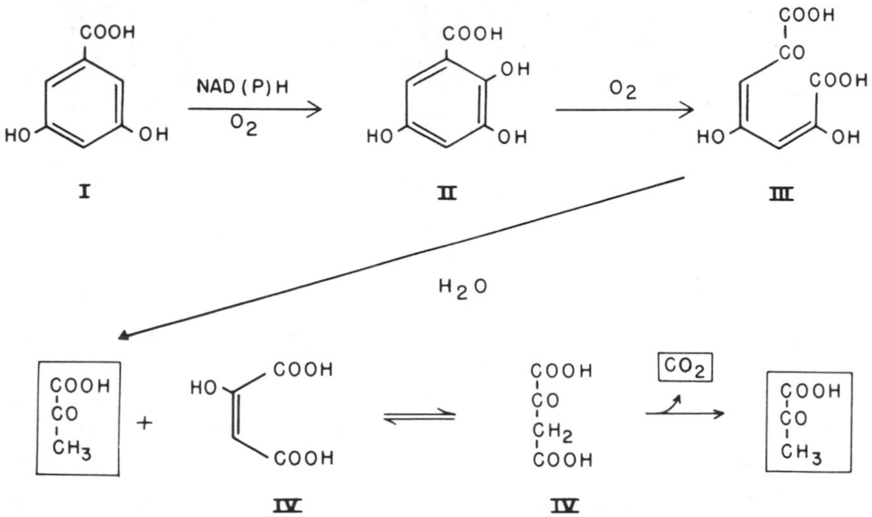

Figure 6.11 Catabolism of 3,5-dihydroxybenzoate by *Bacillus brevis* (Crawford and Olson, 1978a).

Chlorosalicylate Degradation

We have isolated a *Bacillus* strain (5Cl-SAL-1) from the Mississippi River that grows on 5-chloro-2-hydroxybenzoate as its sole source of carbon and energy. The bacterium uses the following dioxygenase to dissimilate chlorosalicylate:

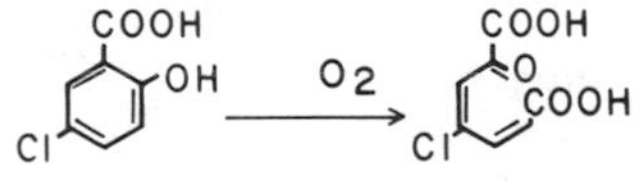

No cofactors are required for oxidation of 5-chlorosalicylate by crude cell-free extracts prepared from chlorosalicylate-grown *Bacillus* 5-Cl-SAL-1. Each mole of 5-chlorosalicylate elicits consumption of one mole of O_2, in the manner of a reaction catalyzed by a dioxygenase. We have purified the oxygenase activity using a combination of ammonium sulfate fractionation and affinity chromatography. The ring-fission product produced from 5-chlorosalicylate shows a $\lambda_{max} = 290$ nm at pH 7 or $\lambda_{max} = 300$ nm at pH 1. Strong base converts the ring-fission product to maleylpyruvate. These and other observations are consistent with the reactions shown in Figure 6.12.

Crude cell-free extracts of chlorosalicylate-grown *B. brevis* contain maleylpyruvate hydrolase, maleylpyruvate isomerase (GSH-dependent), and fumarylpyruvate hydrolase activities. Such extracts do not oxidize gentisate. They convert one mole of 5-chlorosalicylate to one mole of pyruvate and one mole of chloride ion. The 5-chlorosalicylate dioxygenase shows a classic "*meta*"-dioxygenase-like requirement for ferrous ions.

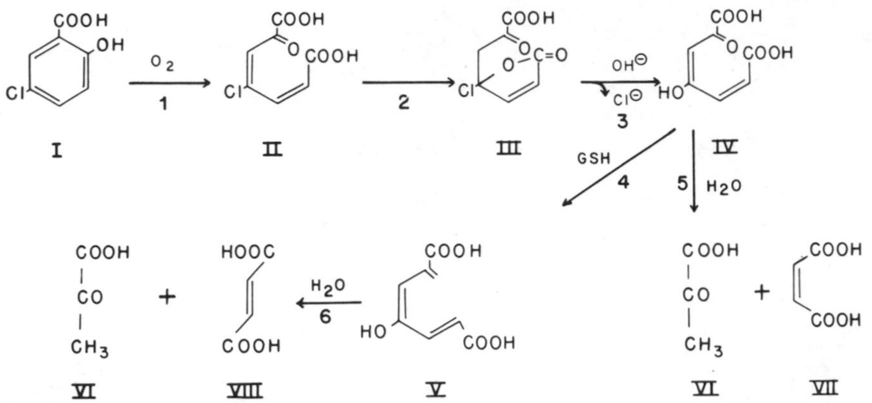

Figure 6.12 Catabolism of 5-chlorosalicylate by a *Bacillus* (R. L. Crawford *et al.*, 1979).

This is apparently the first observation of a specific dioxygenase that cleaves as the enzyme's primary substrate a single-ring benzenoid molecule having only one hydroxyl on the ring. Que (1978) reported that a catechol 2,3-dioxygenase would also cleave the ring of o-aminophenol; however, the reaction rate was very slow in comparison with the rate of cleavage of the ring of the enzyme's normal substrate, catachol. Kiyohara and Nadao (1977) reported the existence of a 1-hydroxy-2-naphthoate 1,2-dioxygenase, where a biphenyl ring structure was oxidatively cleaved while bearing only one hydroxyl group. In the instance discussed here, the monohydroxylated molecule (5-chlorosalicylate) is the primary substrate, while the corresponding dihydroxy compound (gentisate) is not attacked at a significant rate.

6.1.3 Degradation of Benzenoid Compounds by Fungi

Most of the biochemistry discussed in the preceding section was learned by studying the catabolic mechanisms of aromatic compound-degrading bacteria. Much less work has been performed with the fungi, despite their well-known importance as degraders of benzenoid molecules within natural environments. The following briefly summarizes some of the important work that has been done with fungi concerning biochemical mechanisms of degradation of benzenoid molecules.

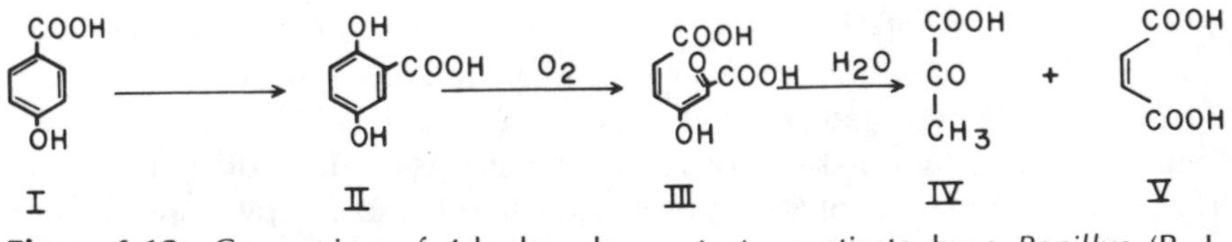

Figure 6.13 Conversion of 4-hydroxybenzoate to gentisate by a *Bacillus* (R. L. Crawford, 1976a).

Only a limited number of the dioxygenases shown in Figure 6.8 have been described in fungi. Those dioxygenases that have been observed in fungi are listed in Table 6.2.

TABLE 6.2 Representative Aromatic Ring-Fission Dioxygenases of Fungi

Fungus	Dioxygenase	Reference
Tilletiopsis washingtonensis	Homoprotocatechuate-3,4	Subba Rao et al. (1971)
Penicillium patulum, Penicillium urticae	2,5-Dihydroxybenzaldehyde-4,5	Scott and Beadling (1974); Forrester and Gaucher (1972)
Aspergillus niger	Homogentisate-1,2	Sugumaran and Vaidyanathan (1978)
Candida tropicalis	4-Methylcatechol-1,6	Hashimoto (1970); Hashimoto (1973)
Tilletiopsis washingtonensis	Protocatechuate-3,4	Subba Rao et al. (1971)
Candida tropicalis	1,3,4,-Trihydroxy-benzene-3,4	Karasevich and Ivoilov (1977)
Candida tropicalis	Catechol-1,2	Neujahr et al. (1974)
Neurospora crassa	Protocatechuate-3,4	Gross et al. (1956)
Aspergillus niger Rhodotorula mucilaginosa Sporobolomyces sp. Penicillium spinulosum Endomycopsis sp. Cylindrocephalum sp. Fusarium oxysporum Polystictus versicolor (=Coriolus versicolor)	Protocatechuate-3,4	Cain et al. (1968)
Debaryomyces subglobosus Rhodotorula mucilaginosa Fusarium oxysporum Penicillium spinulosum Vararia granulosa Schizophyllum commune Debaryomyces subglobosus Debaryomyces hansenii Aureobasidium pullulans	Catechol-1,2	Cain et al. (1968)
Pullaria pullans (=Aureobasidium pullulans)	Protocatechuate-3,4	Henderson (1961)

Fungi are well known for their abilities to decarboxylate aromatic compounds (Ramachandran et al., 1979). Some examples of these abilities are presented in Table 6.3.

TABLE 6.3   Representative Decarboxylations of Benzenoid Compounds by Fungi

Substrate	Product	Organism	Reference
Benzoic acid	Benzene	*Hypoxylum pruinatum* (=*Hypoxylon mammatum*)	Hubbles and Mehram (1968)
Salicylic acid	Phenol	*Glomerella cingulata*	Burwood and Spencer (1970)
2,3-Dihydroxy-benzoic acid	Catechol	*Glomerella cingulata*	Burwood and Spencer (1970)
2,3-Dihydroxy-benzoic acid	Catechol	*Aspergillus niger*	Terui et al. (1953); Terui et al. (1961); Subba Rao et al. (1967)
2,3-Dihydroxy-benzoic acid	Catechol	*Trichoderma lignorum*	Vidal (1969)
2,4-Dihydroxy-benzoic acid	Resorcinol	*Aspergillus nidulans*	Ramanarayanan and Vaidyanathan (1975); Yano and Arima (1958)
Protocatechuic acid	Catechol	*Aspergillus* sp.	Butkevich (1924)
6-Methylsali-cylic acid	*m*-Cresol	*Penicillium patulum*	Light (1969); Light and Vogel (1975)
Orsellinic acid	Orcinol	*Gliocladium roseum*	Pettersson (1965)
Orsellinic acid	Orcinol	*Umbilicaria pustulata*	Mosbach and Ehrensvard (1966)
p-Hydroxy-benzoic acid	Hydro-quinone	*Candida tropicalis*	Karasevich and Ivoilov (1977)
2,3-Dihydroxy-benzoic acid	Catechol	*Aspergillus sojae* (=*A. parasiticus*)	Yuasa et al. (1978a)
3,4-Dihydroxy-cinnamate	Hydroxy-styrene	*Polyporus circinata*	Bayne et al. (1976)
trans-Cinnamic acid	Styrene	*Aspergillus niger*	Herzog and Ripke (1908)
trans-Cinnamic acid	Styrene	*Pencillium* sp.	Jaminet (1950)
Protocatechuate	Catechol	*Debaryomyces* sp.	Cain et al. (1968)

Fungi are also noted for their propensity to hydroxylate aromatic rings or substituents on aromatic rings (Sugumaran and Vaidynathan, 1979). Table 6.4 illustrates some representative examples of such fungal hydroxylation reactions.

TABLE 6.4 Representative Hydroxylations of Benzenoid Molecules by Fungi

Substrate	Product	Organism	Reference
o-Cresol	3-Methyl-catechol	*Candida tropicalis*	Hashimoto (1973); Hashimoto (1970)
Phenol	Catechol	*Trichosporon cutaneum*	Neujahr and Varga (1970); Neujahr and Gaal (1973)
p-Cresol	4-Methyl-catechol	*Candida tropicalis*	Hashimoto (1973)
m-Cresol	4-Methyl-catechol	*Candida tropicalis*	Hashimoto (1973)
m-Cresol	1,4-Dihydroxy-2-methyl-benzene	*Penicillium urticae*	Forrester and Gaucher (1972)
m-Cresol	3-Hydroxy-benzyl alcohol	*Penicillium patulum*	Light (1969); Light and Vogel (1975)
m-Cresol	1,4-Dihydroxy-2-methyl-benzene	*Pencillium patulum*	Light (1969); Light and Vogel (1975)
3-Hydroxy-benzyl alcohol	1,4-Dihydroxy-benzyl alcohol	*Pencillium patulum*	Light (1969); Light and Vogel (1975)
Salicylic acid	Catechol	*Pullularia pullulans (=Aureobasidium pullalans)*	Henderson (1961)
Salicylic acid	2,3-Dihydroxy-benzoic acid	*Aspergillus niger*	Terui et al. (1953)
Salicylic acid	2,3-Dihydroxy-benzoic acid	*Aspergillus nidulans*	Shepherd and Villanueva (1959)
Salicylic acid	2,3-Dihydroxy-benzoic acid	*Trichoderma lignorum*	Vidal (1969)
Salicylic acid	2,4-Dihydroxy-benzoic acid	*Aspergillus nidulans*	Ramanarayanan and Vaidyanathan (1975)
Salicylic acid	2,4-Dihydroxy-benzoic acid	*Trichosporon sp.*	Fujikawa and Ito (1971) [cited in Harda and Watanabe (1972)]

TABLE 6.4 (Continued)

Substrate	Product	Organism	Reference
Salicylic acid	2,5-Dihydroxy-benzoic acid	*Trichoderma* sp.	Harda and Watanabe (1972)
Salicylic acid	2,6-Dihydroxy-benzoic acid	*Trichosporon* sp.	Fujikawa and Ito (1971)
Benzoic acid	3-Hydroxy-benzoic acid	*Aspergillus niger*	Bocks (1967a)
Vanillic acid	Methoxyhydro-quinone	*Polyporus dichrous* (=*Gloeoporus dichrous*)	Kirk and Lorenz (1974)
Phenylacetic acid	3-Hydroxyphenyl-acetic acid	*Aspergillus niger*	Faulkner and Woodcock (1968)
Phenylacetic acid	2-Hydroxy-phenylacetic acid	*Pencillium chrysogenum*	Nishikada (1951) [cited in Sugumaran and Vaidyanathan (1978)]
Phenylacetic acid	2-Hydroxy-phenylacetic acid	*Aspergillus sojae* (=*A. parasiticus*)	Yuasa *et al.* (1975)
Phenylacetic acid	2-Hydroxy-phenylacetic acid	*Schizophyllum commune*	Moore and Towers (1967)
Phenylacetic acid	2-Hydroxy-phenylacetic acid	*Alternaria*	Kohmoto *et al.* (1970)
Benzoic acid	4-Hydroxy-benzoic acid	*Aspergillus niger*	Reddy and Vaidyanathan (1975)
Phenylacetic acid	2-Hydroxy-phenylacetic acid	*Fusarium* sp.	Kunita (1955)
Phenylacetic acid*	3-Hydroxy-phenylacetic acid	*Rhizoctonia* salani	Kohmoto *et al.* (1970)
Phenylacetic acid	2-Hydroxy-phenylacetic acid	*Aspergillus niger*	Faulkner and Woodcock (1968)
Phenylacetic acid	3-Hydroxy-phenylacetic acid	*Pencillium* sp.	Kunita (1955)
Phenylacetic acid	2-Hydroxy-phenylacetic acid	*Pencillium chrysogenum*	Isono (1958)

TABLE 6.4 (Continued)

Substrate	Product	Organism	Reference
4-Hydroxy-phenylacetic acid	3,4-Dihydroxy-phenylacetic acid	*Pencillium chrysogenum*	Isono (1958)
Homogentisic acid	2,5-Dihydroxy-mandelic acid	*Polyporus tumulosus*	Fujikawa and Ito (1971)
Cinnamic acid	2-Hydroxy-cinnamic acid	*Aspergillus niger*	Bocks (1967)
Cinnamic acid	4-Hydroxy-cinnamic acid	*Rhizoctonia solani*	Kalghatgi *et al.* (1974)
Cinnamic acid	4-Hydroxy-cinnamic acid	*Polystictus versicolor*	Farmer *et al.* (1959)
p-Coumaric acid	Caffeic acid	*Lentinus lepideus*	Power *et al.* (1965)
p-Coumaric acid	3,4-Dihydroxy-cinnamic acid	*Lentinus lepideus*	Farmer *et al.* (1959)
Hydroquinone	1,3,4-Tri-hydroxybenzene	*Candida tropicalis*	Karasevich and Ivoilov (1977)
Tyrosine	2,3-Dihydroxy-phenylalanine	*Aspergillus sojae (=A. parasiticus)*	Yuasa *et al.* (1978)
Naphthalene	1,2-Naphthalene oxide	*Cunninghamella elegans*	Cerniglia *et al.* (1977)
Naphthalene	1-Naphthol via 1,2-naphthalene oxide	47 species in 34 genera	Cerniglia *et al.* (1978)
Various 4-alkoxybenzoic acids	Various hydroxy-benzoic acids	*Polyporus dichrous*	Kirk and Lorenz (1974)
Phenylacetic acid	Homogentisic acid	*Aspergillus niger*	Kishore *et al.* (1976)
Phenylacetic acid	2,6-Dihydroxy-phenylacetic acid	*Aspergillus fumigatus*	Yoshizako *et al.* (1977)
4-Hydroxy-benzoic acid	Protocatechuic acid	Nine species of various genera	Cain *et al.* (1968)
m-Cresol	2,5-Dihydroxy-toluene	*Pencillium patulum*	Murphy *et al.* (1974); Scott *et al.* (1973)

Fungal transformations of benzenoid molecules are often associated with processes that have been lumped under the term "secondary metabolism." In fact, lignin biodegradation by certain white-rot fungi is probably a secondary metabolic process (Kirk et al., 1978; Section 6.4). Thus it is often difficult to demonstrate a "purpose" for many fungal transformations of aromatic compounds. Most of the research summarized in Tables 6.2–6.4 involves the isolation of small quantities of fungal transformation products of benzenoid molecules. The fungi often do not utilize the aromatic compounds as sole carbon and energy sources, but transform them while using alternative energy sources such as glucose or complex media such as potato–dextrose broth (Cerniglia et al., 1978). Some fungi, however, do use aromatic compounds as sole sources of carbon and energy (Karasevich and Ivoilov, 1975; Henderson and Farmer, 1955; Ohta et al., 1979). In such cases degradation proceeds ultimately to CO_2, H_2O, and cell mass.

6.2 MICROBIAL CATABOLISM OF LIGNIN MODEL COMPOUNDS

It is an exceedingly difficult task to study the biochemical mechanisms whereby microorganisms degrade lignin. A primary reason for this difficulty is the pronounced structural complexity of the lignin molecule (for recent schematic representations of spruce and beech lignins see Figures 2.3 and 2.4, respectively). Unlike other biopolymers, lignin contains no readily hydrolyzable bond recurring at periodic intervals along a linear backbone. Instead, lignin is a three-dimensional, amorphous polymer containing many different stable carbon-carbon and ether linkages between phenylpropanoid monomeric units. Thus it is difficult to design experiments to ascertain what specific enzymic transformations are occurring during microbial decay of lignin. Usually, at best gross chemical alterations (e.g., by quantification of various functional groups) in the lignin polymer before and after decay may be measured (Kirk and Chang, 1974).

Theoretically, one way to circumvent this problem of chemical complexity is to study the microbial degradation of simple lignin model compounds of known chemical structure. In this experimental approach low-molecular-weight compounds that contain chemical structures known to occur in lignin are used (Figure 6.14). It is then assumed that what may be learned concerning the biochemistry of degradation of these lignin models is relevant to catabolic mechanisms of lignin biodegradation (Muranaka et al., 1976). As the following discussion indicates, it is not yet clear whether this is a valid experimental approach.

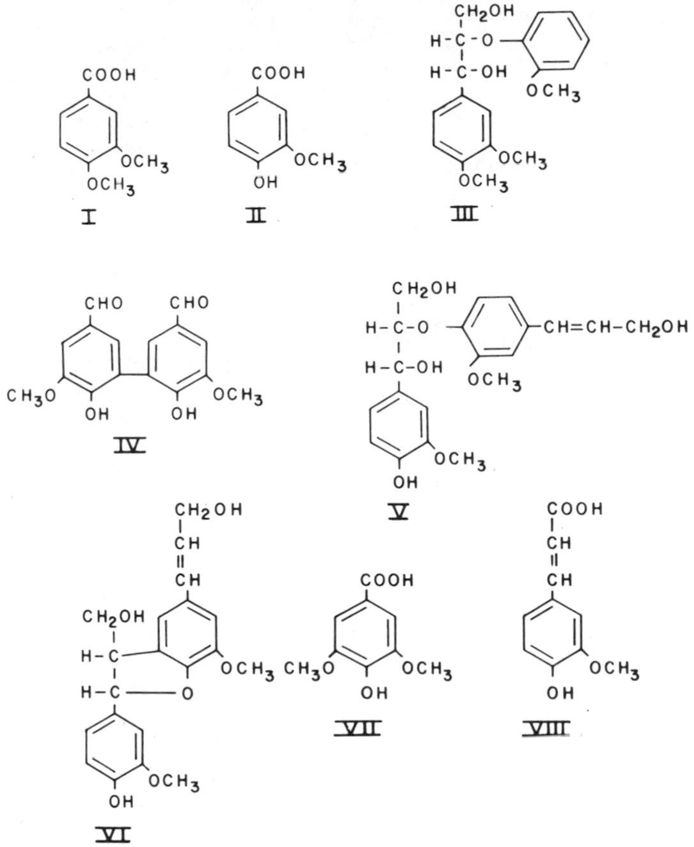

Figure 6.14 Some representative lignin model compounds: (I) veratric acid; (II) vanillic acid; (III) veratrylglycerol-β-(o-methoxyphenyl) ether; (IV) dehydrodivanillin; (V) guaiacylglycerol-β-coniferyl alcohol ether; (VI) dehydrodiconiferyl alcohol; (VII) syringic acid; (VIII) ferulic acid (Crawford and Crawford, 1980).

6.2.1 Methoxylated Aromatic Acids as Lignin Model Compounds

The simple methoxylated aromatic acids veratric acid (3,4-dimethoxybenzoic acid), vanillic acid (4-hydroxy-3-methoxybenzoic acid), and syringic acid (3,4-dimethoxy-4-hydroxybenzoic acid) have been used frequently as model substrates for the isolation of microorganisms with the intent of studying lignin biodegradation mechanisms (Figure 6.15 shows the relationship between structural elements of hardwood lignin and veratric acid and syringic acid). It is usually assumed (Kawakami, 1976a,b; Henderson and Farmer, 1955) that knowledge gained concerning the mechanisms

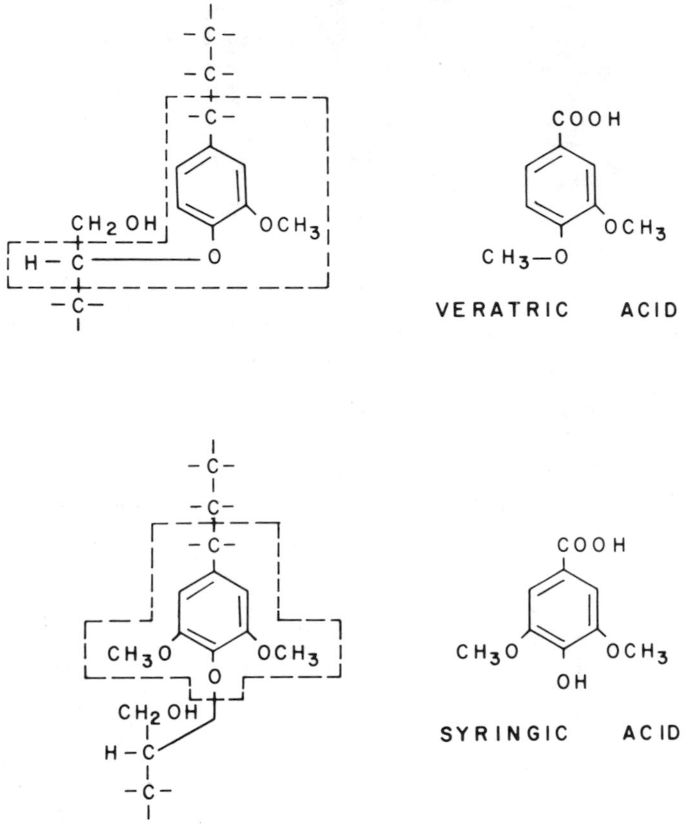

Figure 6.15 The relationship between structural elements of lignin and simple methoxylated aromatic acids (Crawford and Crawford, 1980).

of microbial demethylation and ring fission of vanillate, veratrate, and/or syringate is relevant to lignin biodegradation in that the same types of degradative transformations (Kirk and Chang, 1974) are suspected to occur during lignin biodegradation.

Vanillate and veratrate are usually demethylated by microorganisms, yielding protocatechuate (3,4-dihydroxybenzoate), which then undergoes dioxygenase-catalyzed ring fission (Flaig and Haider, 1961; R. L. Crawford et al., 1973; R. L. Crawford, 1975a; Cartwright and Smith, 1967; Cartwright and Buswell, 1967; Cain et al., 1968; Toms and Wood, 1969; Tadasa, 1977; Kawakami, 1976a). Kirk and Lorenz (1973) described an unusual oxidative decarboxylation of vanillate to methoxyhydroquinone by whole cells of the wood-rotting fungus *Polyporus dichrous* (=*Gloeoporus dichrous*), though further catabolism of this potential ring-fission substrate was not observed. Crawford and Olson (1978b) reported that several strains of *Bacillus megaterium* and one strain of a *Streptomyces* sp. were able to nonoxidatively decarboxylate vanillate, yielding *o*-methoxyphenol and CO_2 (veratrate was also converted to *o*-methoxyphenol, probably by way of vanillate). The known transformations of vanillate and veratrate are summarized in Figure 6.16.

In a recent study Buswell et al. (1979) partially purified an oxidative vanillate decarboxylase from the white-rot fungus *Phanerochaete chrysosporium* (=*Sporotrichum pulverulentum*). The decarboxylase was dependent on NADPH or NADH as coenzyme, and its activity was slightly stimulated by addition of FAD to reaction mixtures. Decarboxylase activity was inducible in the *Phanerochaete* by the presence of vanillate in the growth medium. Enzyme activity was inhibited by the Fe^{3+} chelator Tiron, but not by the Fe^{2+} chelator α,α-dipyridyl or by EDTA. The decarboxylase was highly nonspecific, oxidatively decarboxylating numerous 4-hydroxybenzoate derivatives other than vanillate (e.g., protocatechuate, gallate, and 4-hydroxybenzoate). The decarboxylase is an intracellular enzyme. This observation, along with the fact that the decarboxylase appears prior to nitrogen starvation conditions (Section 6.4), indicates that this enzyme probably is not directly involved as a component of the early steps in lignin

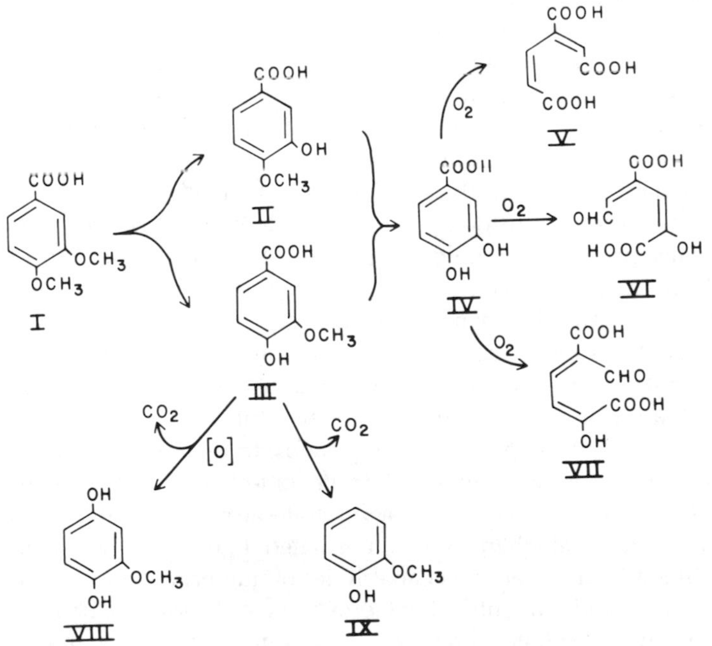

Figure 6.16 Microbial transformations of veratric acid and vanillic acid (Crawford and Crawford, 1980).

biodegradation, even though it may be involved in degradation of low-molecular-weight lignin-derived phenolics. Yajima et al. (1979) have also studied the vanillate hydroxylase of *Phanerochaete chrysosporium*, confirming and extending many of the observations of Buswell et al. (1979).

In another study Buswell and Eriksson (1979) demonstrated a hydroxyquinol (1,2,4-trihydroxybenzene) 1,2-dioxygenase in crude cell-free extracts prepared from mycelial mats of glucose-grown *Sporotrichum pulverulentum* (=*P. chrysosporium*). The dioxygenase was induced threefold by the presence of vanillate in the growth medium; thus it appears likely that in *Phanerochaete chrysosporium* vanillate is metabolized by the sequence vanillate $\rightarrow$ methoxyhydroquinone $+ CO_2$ $\rightarrow\rightarrow$ hydroxyquinol $\rightarrow\rightarrow$ maleylacetate. *Sporotrichum pulverulentum* also produces an intracellular quinone oxidoreductase system which reduces methoxyquinone to methoxyhydroquinone (Buswell et al., 1979). The reductase is induced during growth in the presence of vanillate. It is possible that this reductase's function is to reverse any enzymic or nonenzymic conversion of methoxyhydroquinone to methoxyquinone to ensure complete conversion of vanillate to Krebs cycle intermediates (Buswell et al., 1979).

Ander et al. (1980) examined vanillic acid degradation by *Sporotrichum pulverulentum*. These authors confirmed the occurrence of oxidative decarboxylation of vanillic acid to methoxyhydroquinone by *S. pulverulentum*, but also observed an alternative pathway involving reduction of vanillic acid to vanillin and vanillyl alcohol [Ishikawa et al. (1963) observed similar reductions of vanillic acid by strains of *Polyporus* and *Fomes*]. The reductive pathway occurred only in the presence of an easily metabolized carbon source such as glucose. By using radiolabeled substrates it was observed that in the oxidative pathway decarboxylation of vanillic acid preceded ring cleavage, which in turn took place before demethylation of the methoxyl group. This observation indicates that the true ring-fission substrate for catabolism of vanillic acid by *S. pulverulentum* may be a trihydroxymethoxybenzene rather than hydroxyquinol.

It is not yet possible to decide whether the study of the microbial degradation of vanillate and/or veratrate is significantly relevant to the study of lignin biodegradation. In fact, it appears that the ability of bacteria (isolated for their ability to degrade phenolic compounds) to degrade vanillate does not necessarily correlate with an ability to degrade lignin. R. L. Crawford et al. (1973) showed that whole cells of a veratrate-grown *Nocardia* produced no detectable structural changes in an isolated lignin. Haider et al. (1978) examined the lignin-degrading capabilities of numerous strains of *Nocardia*, *Pseudomonas*, and unidentified bacteria that were known to be degraders of phenols such as vanillate, veratrate, and anisate (methoxybenzoate). Several of the unidentified bacteria were isolated from lakewater polluted by pulpmill waste lignin. Five of the lakewater strains were shown to grow on vanillate; however, none were able to release more than 3.8% of the ^{14}C in O^{14}-CH_3-DHP-lignin after 15 days of incubation. Several strains of *Nocardia* (obtained from the German Collection of Microorganisms) were tested for their abilities to degrade vanillate and $O^{14}CH_3$-DHP-lignin. There was apparently no significant correlation between these two abilities. Some strains that readily degraded vanillate did not readily demethylate the DHP-lignins, while other strains that degraded vanillate more slowly demethylated the DHP-lignins quite well. Numerous vanillate-degrading *Pseudomonas* species released only small amounts of CO_2 from the methoxyl groups of DHP-lignins. All the bacteria examined in this study were unable to evolve significant amounts of $^{14}CO_2$ from ^{14}C-[lignin-labeled]-lignocelluloses prepared from corn stalks.

It is not at all surprising that vanillate-, veratrate-, and syringate-degrading microorganisms do not necessarily degrade lignin. Microorganisms catabolize these compounds by way of highly substrate-specific pathways such as the β-ketoadipate pathway (Ornston and Stanier, 1973), the protocatechuate 2,3-dioxygenase pathway (R. L. Crawford, 1975a), or the 3-0-methylgallate pathway (Sparnins and Dagley, 1975). It is highly unlikely that such highly specific pathways also function in degradation of the lignin macromolecule. These pathways, however, might function in the catabolism of vanillate, veratrate, or syringate themselves if these compounds or their precursors were released from lignin as degradative fragments following decay of the lignin polymer by other enzyme systems (Ishikawa et al., 1963; Haars and Hüttermann, 1980). It is probable that the same *types* of enzymes (mixed-function oxygenases and dioxygenases; Dagley, 1971) that mediate degradation of low-molecular-weight aromatic molecules are involved in lignin biodegradation (Kirk and Chang, 1974), but this has not been unequivocally established.

Chen et al. (1979) have presented preliminary evidence to show that *ortho* (intradiol) ring-fission reactions occur during degradation of lignin by *Phanerochaete chrysosporium*. The occurrence of such ring-fission reactions was established by determining the chemical structures of about 30 lignin decay fragments.

An extensive survey of the vanillate-degrading abilities of microorganisms known to be lignin degraders [the reverse of the screening experiments of Haider et al. (1978)] has apparently not been undertaken. In the few instances where such a relationship has been examined, lignin-degrading microorganisms have been shown also to be degraders of vanillate (Haider and Grabbe, 1967; Gradziel et al., 1978) or other simple lignin model compounds (Haider and Trojanowski, 1975). Thus more research may show that the correlation between model compound utilization and lignin degradation is not that model compound-degraders decompose lignin, but that lignin-degraders are able to decompose lignin models. However, Kirk and

Lorenz (1973) found that the lignin-degrading fungus *Polyporus dichrous* (=*Gloeoporus dichrous*) did not degrade syringic acid, despite the fact that this fungus is a degrader of hardwood lignins that contain appreciable amounts of syringyl structures. As Ander and Eriksson (1978) point out, this discrepancy shows that catabolism of lignin in wood may proceed by different mechanisms than catabolism of lignin model substrates in liquid cultures.

6.2.2   Veratrylglycerol-β-(o-methoxyphenyl) Ether and Related Lignin Model Compounds

Several papers have been published establishing the abilities of microorganisms to degrade lignin model compounds that contain the arylglycerol-β-aryl ether structure (Kawakami, 1975a,b; Trojanowski et al., 1970; R. L. Crawford et al., 1975; R. L. Crawford et al., 1973b; Fukuzumi et al., 1969; Weinstein et al., 1980; Enoki et al., 1980). This linkage type represents 30–50% of the intermonomer bonds in spruce lignin (Adler, 1977; see compounds III and V of Figure 6.14). R. L. Crawford et al. (1973b, 1975) mapped the catabolic pathway used by *Pseudomonas acidovorans* to degrade veratrylglycerol-β-(o-methoxyphenyl) ether, as shown in Figure 6.17. The pseudomonad was able to completely degrade both rings and the propanoid side chain of this lignin model. The *Pseudomonas*, however, was unable to release more than a few percent of the ^{14}C in ^{14}C-[lignin-labeled]-spruce as $^{14}CO_2$ over an incubation period of several weeks (R. L. Crawford, unpublished results). Thus a correlation between lignin model compound degradation and lignin degradation is again unsupported. Fukuzumi et al. (1969) also observed oxidative cleavage of the arylglycerol-β-aryl ether bond of veratrylglycerol-β-(o-methoxyphenyl) ether; however, these authors used an enzyme preparation from the white-rot fungus *Poria subacida* (=*Perenniopoia subacida*). Again, an implication of this work is that the correlation between model compound degradation and lignin degradation is that lignin-degrading microorganisms are able to degrade lignin models, but model degraders are not necessarily degraders of lignin.

Fukuzumi and Katayama (1977) studied the degradation of guaiacylglycerol-β-coniferyl ether by a strain of *Pseudomonas putida*, which used the lignin model as a sole source of carbon and energy. Isolation of degradation fragments allowed the authors to postulate the catabolic reaction shown in Figure 6.18. Since no enzymatic studies were undertaken by the authors, it is not clear how many reactions (enzymatic and/or nonenzymatic) are involved in the conversion of guaiacylglycerol-β-coniferyl ether to β-hydroxypropiovanillone and coniferyl alcohol. Also unresolved is the mechanism whereby the dimer was split to form these two monomeric units.

Krisnagakura et al. (1979) have reported in a preliminary communication

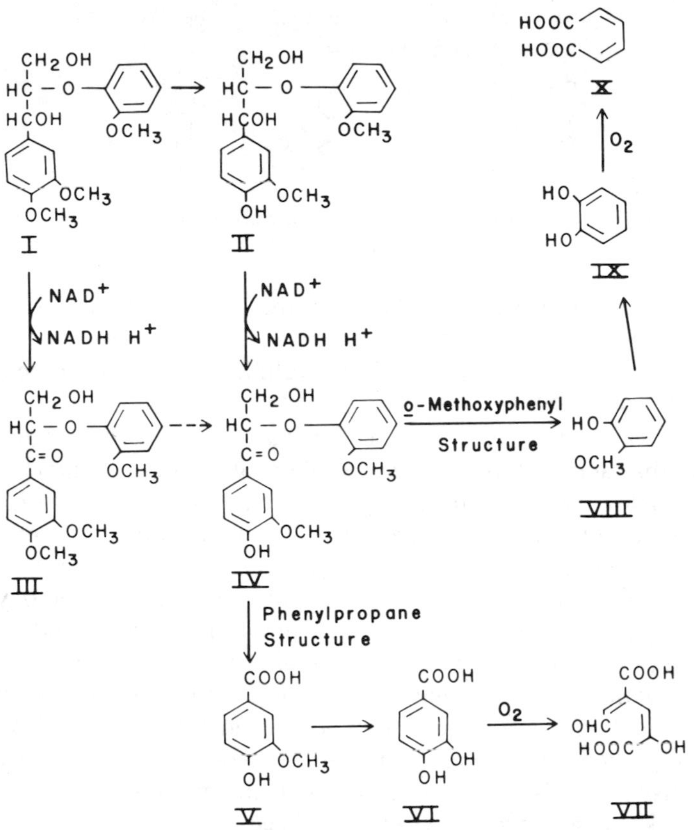

Figure 6.17 Bacterial degradation of the lignin model compound vertarylglycerol-β-(o-methoxyphenyl) ether. The conversions of compound I to compound III, and II to IV may have some relevance to degradation of the lignin polymer in that it is known that white-rotted lignin contains increased amounts of benzylketo groups as compared to nondecayed lignin (Kirk and Chang, 1974). The enzymes of the above pathway, however, are intracellular and would probably not attack the intact lignin macromolecule (see text; Crawford and Crawford, 1980).

that the white-rot fungus *Phanerochaete chrysosporium* degrades ^{14}C-labeled veratrylglycerol-β-guaiacyl ether (compound I, Figure 6.17) to $^{14}CO_2$, and that this transformation is apparently under the same types of metabolic control mechanisms as are the enzymes that convert ^{14}C-lignin to $^{14}CO_2$. For example, lag phases for both conversions are similar, and both conversions are repressed by addition of nitrogen to the growth medium (Keyser *et al.*, 1978). These results indicate that degradation of veratrylglycerol-β-guaiacyl ether and lignin are closely related phenomena. Weinstein *et al.* (1980) have shown that ^{14}C-labeled guaiacylglycerol-β-

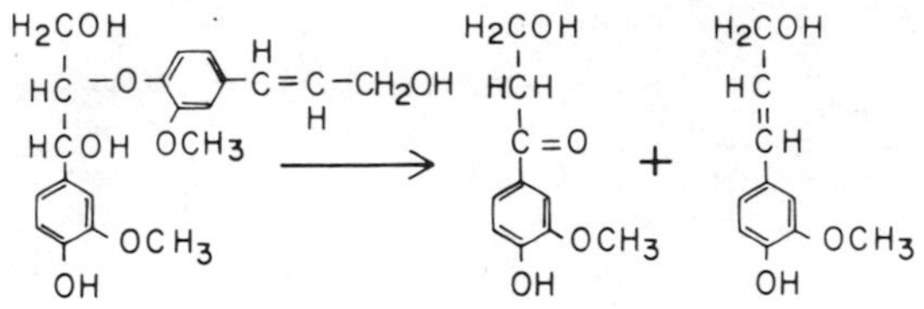

$$H_2COH \qquad\qquad H_2COH \qquad H_2COH$$

$$HC-O-\text{(ring)}-C=C-CH_2OH \quad HCH \qquad HC$$

$$HCOH \quad OCH_3 \longrightarrow \quad C=O \quad + \quad CH$$

(with OCH$_3$/OH substituted aromatic rings)

Figure 6.18 Biotransformation of guaiacylglycerol-β-coniferyl ether by *Pseudomonas putida* (Fukuzumi and Katayama, 1977).

guaiacyl ether is converted to $^{14}CO_2$ 2–3 times faster when 100% O_2 rather than air (21% O_2) is present above cultures of *Phanerochaete chrysosporium*. This again is similar to $^{14}CO_2$ evolution patterns observed during lignin degradation by this organism.

Enoki *et al.* (1980) examined the metabolism of veratrylglycerol-β-guaiacyl ether and 4-ethoxy-3-methoxyphenylglycerol-β-guaiacyl ether by *Phanerochaete chrysosporium*. 2-(*o*-Methoxyphenoxy)ethanol was identified as a product of the catabolism of both lignin models. Veratryl alcohol was produced from veratrylglycerol-β-guaiacyl ether, while 4-ethoxy-3-methoxybenzyl alcohol was produced from 4-ethoxy-3-me-thoxy-phenylglycerol-β-guaiacyl ether. These observations indicate that *P. chrysosporium* produces an enzyme or enzymes capable of cleaving the α,β bonds of these dimeric lignin model compounds. An implication of this work is that *P. chrysosporium's* ligninolytic enzyme system may cleave lignin's important arylglycerol-β-aryl ether linkages by fission of the α,β carbon-carbon bonds of the glycerol side chains, rather than by direct oxidative cleavage of the β-aryl ether linkages.

6.2.3 An Alcohol Oxidase That Oxidizes Lignin

Iwahara and Higuchi (1979) have presented preliminary evidence to show that an alcohol oxidase purified from culture filtrates of the fungus *Fusarium solani* after growth in the presence of dehydropolymers (Ohta *et al.*, 1979) nonspecifically oxidizes α,β-unsaturated primary alcohols (e.g., coniferyl alcohol) with production of the corresponding aldehyde and H_2O_2. The oxidase attacks a wide range of lignin model compounds containing α,β-unsaturated primary alcohol structures. It also attacks the macromolecular structure of various lignin preparations. Though these results have been presented only in preliminary form, they are potentially of great significance. This is the first unequivocal demonstration of attack by a purified enzyme (other than laccase- or peroxidase-type phenol oxidases) on the lignin macromolecule. The reation catalyzed by the enzyme apparently prepares phenylpropanoid side chains of lignin monomeric units for further oxidation to the level of C_6–C_1 structures (types of degradative reactions already known to occur during white-rot of lignin; Hata, 1966). The enzyme is extracellular and highly nonspecific in its attack on lignin model compounds, attributes expected of an enzyme involved in degradation of the lignin macromolecule. The work of Iwahara and Higuchi (1979) is the best indication yet that the use of lignin model compounds may under some circumstances have great relevance to the study of biodegradation of the lignin polymer.

6.2.4 Biodegradation of Other Lignin Model Compounds

Katayama and Fukuzumi (1978) studied the bacterial degradation of dehydrodiconiferyl alcohol as a model for coumarin ring structures of lignin. A strain of *Pseudomonas putida* was shown to utilize this dimeric model as a sole source of carbon and energy. Identification of catabolites such as coniferyl alcohol, ferulic acid, and 7-methoxy-3-hydroxymethyl-5(2"-carboxyvinyl)-2-(4-hydroxymethyl-3'-methoxyphenyl)-2,3-dihydrobenzofuran allowed postulation of the catabolic sequence shown in Figure 6.19. Again, potentially valuable enzymological studies to confirm this sequence were apparently not performed. Any relevance of this pathway to lignin biodegradation mechanisms is speculative.

In another study Katayama and Fukuzumi (1979) examined the bacterial degradation of α-veratryl-β-guaiacylpropionic acid and D,L-pinoresinol by *Pseudomonas putida*. The former compound was used as a model for 1,2-diarylpropane structures of lignin. Again, only product isolation studies

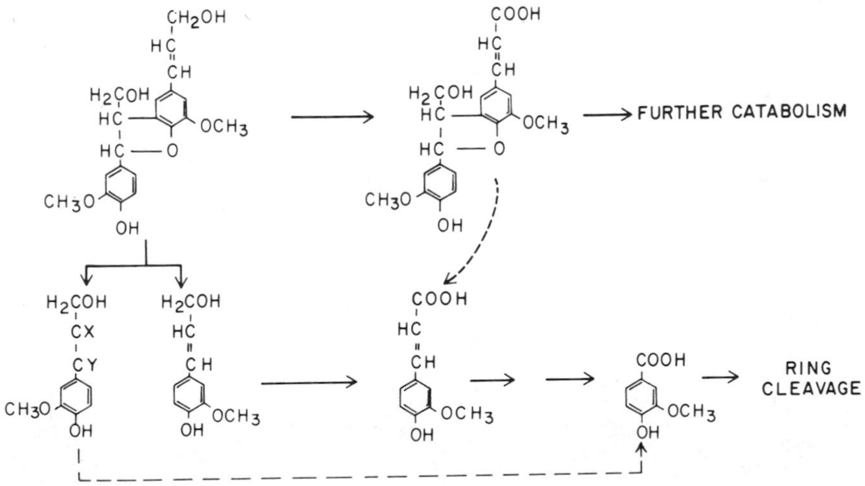

Figure 6.19 Catabolism of dehydrodiconiferyl alcohol by *Pseudomonas putida* (Katayama and Fukuzumi, 1978).

were reported. α-Veratryl-β-guaiacylpropionic acid was converted to verat-raldehyde, vanillin, veratric acid, vanillic acid, and α,β-diguaiacylpropionic acid. Pinoresinol was converted to vanillic acid. From these results a catabolic pathway (Figure 6.20) was proposed for degradation of veratryl-guaiacylpropionate by the pseudomonad.

Konetzka et al. (1952) and Tabak et al. (1959) elucidated catabolic path-ways for the degradation of α-conidendrin by a Flavobacterium and a Pseudomonas, respectively. Both research groups postulated the same se-quence, shown in Figure 6.21. Sundman (1962) studied the degradation of α-conidendrin by an unidentified Gram-negative bacterium, finding a differ-ent catabolic sequence (Figure 6.22) involving isovanillic acid rather than vanillic acid. In both sequences, however, protocatechuic acid appeared to be a key intermediate.

α-Conidendrin is a so-called "lignan" which occurs in small amounts in sprucewood (Freudenberg and Knof, 1957). Its relevance to lignin structure is marginal, since such dimeric structures do not occur in lignin. There is a slight structural similarity between pinoresinol structures in lignin and α-coni-dendrin.

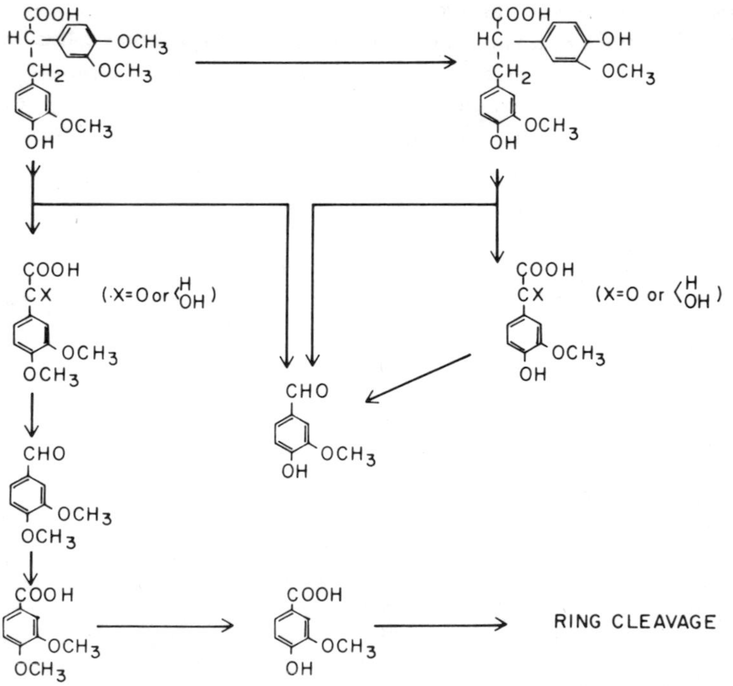

Figure 6.20 Degradation of a α-veratryl-β-guaiacylpropionic acid by Pseudomonas putida (Katayama and Fukuzumi, 1979).

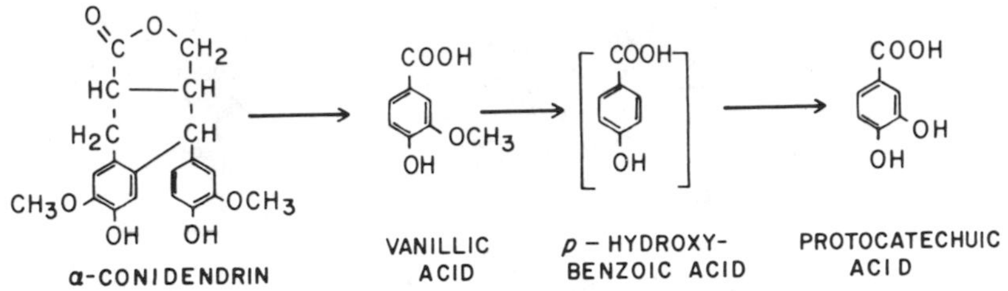

Figure 6.21 Catabolism of α-conidendrin by a *Flavobacterium* and a *Pseudomonas* (Konetzka et al., 1952; Tabak *et al.*, 1959).

The demethoxylations postulated in Figures 6.21 and 6.22 are highly unusual reactions, particularly for bacteria. Bacteria usually convert vanillate or isovanillate to protocatechuate by demethylation of the methoxyl group (e.g., R. L. Crawford *et al.*, 1973), rather than by demethoxylation and hydroxylation as suggested in the pathways of Figures 6.21 and 6.22. A critical examination of the data presented by Konetzka *et al.* (1952), Tabak *et al.* (1959), and Sundman (1962) leads to the conclusion that the postulated demethoxylations of vanillate and isovanillate are not wholly supported by the provided data (see Toms and Wood, 1970). It is possible that the conversions of vanillate and isovanillate to protocatechuate are a result of microbial demethylation reactions rather than demethoxylation and rehydroxylation of the rings. It should be pointed out that microbial demethylations of aromatic methoxyl groups were not well characterized until about 1967 (Cartwright and Smith, 1967; Ribbons and Ohta, 1970), several years after the α-conidendrin studies discussed here.

Toms and Wood (1970) also examined the bacterial degradation of α-conidendrin. A strain of *Pseudomonas multivorans* was isolated by elective culture using α-conidendrin as a sole carbon and energy source. Two new metabolites of α-conidendrin were isolated from culture filtrates, allowing postulation of a reaction sequence for the initial steps in the oxidative degradation of α-conidendrin by the pseudomonad (Figure 6.23). Toms and Wood

see next p.

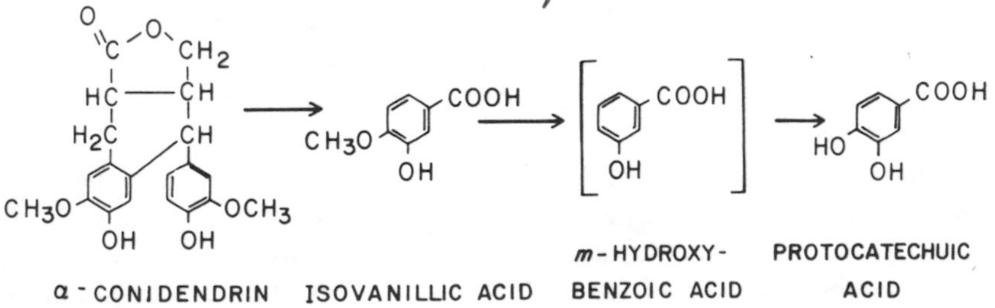

Figure 6.22 Catabolism of α-conidendrin by an unidentified soil bacterium (Sundman, 1962).

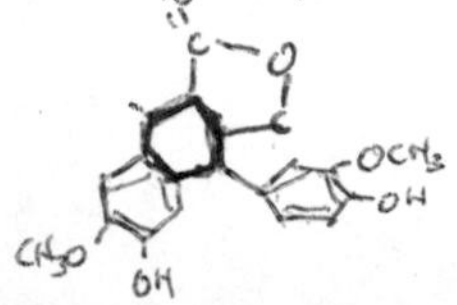

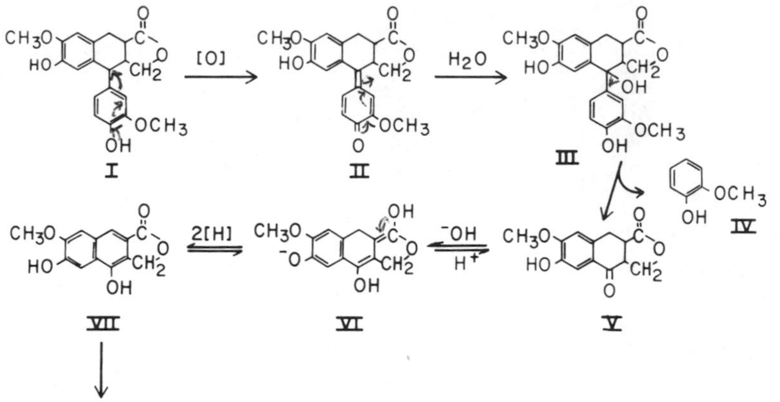

Figure 6.23   Initial reactions in the oxidative degration of α-conidendrin (Toms and Wood, 1970). Reprinted with permission from *Biochemistry*. **9:** 733–740. Copyright by the American Chemical Society. Compounds V and VI were isolated from culture filtrates of *Pseudomonas multivorans* following growth on α-conidendrin as a sole source of carbon and energy.

(1970) postulated that further catabolism of compound VII (Figure 6.23) would probably proceed by way of vanillate and protocatechuate, though no evidence was presented to rule out the possible intermediacy of isovanillate.

Ohta *et al.* (1979) isolated a strain of *Fusarium solani* that utilized DHP lignin as a role source of carbon and energy. The authors then used this organism to study biodegradation of the lignin model dehydrodiconiferyl alcohol. Six metabolic products were isolated and identified, allowing postulation of the catabolic sequence shown in Figure 6.24.

Crawford and Crawford (unpublished results) have found that certain lignin-degrading strains of *Streptomyces* are also able to degrade the lignin model compound dehydrodivanillin. When mutants were produced by irradiation of streptomycete spores with UV light, numerous mutant strains that had lost the ability to attack dehydrovanillin were obtained. These "dehydrodivanillin negative" mutants still retained an unchanged ability to degrade ^{14}C-lignins to ^{14}CO$_2$. This clearly demonstrates that in this instance dehydrodivanillin is not a relevant model for the study of polymeric lignin biodegradation.

Watkins (1970) studied the bacterial degradation of lignosulfonates and various related model compounds. Using soil bacteria as biodegradative agents, he found a relationship between structure and biodegradability of lignosulfonate models. Sulfonate substitution on side chains (e.g., vanillin vs. vanillylsulfonate) greatly reduced biodegradability of the lignin models.

Healy and Young (1979) have shown that a range of 11 simple aromatic

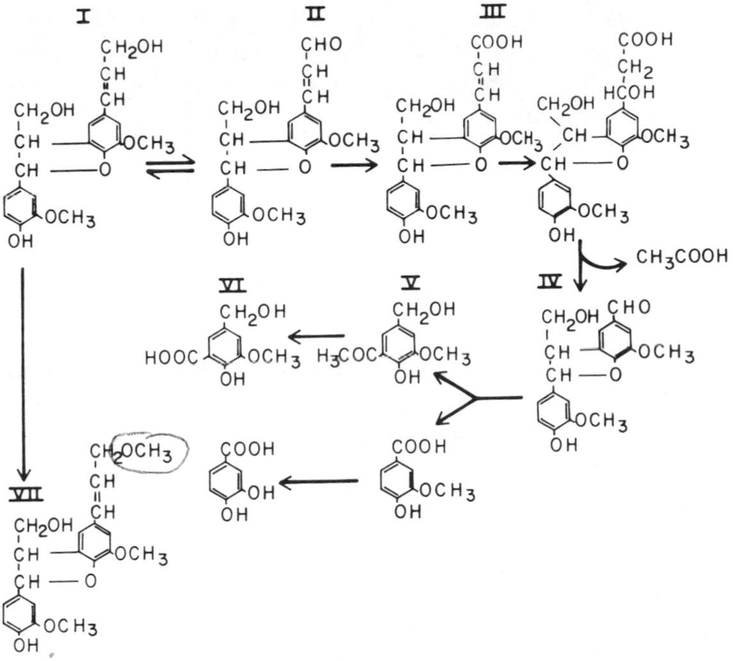

Figure 6.24 Degradation of dehydrodiconiferyl alcohol by *Fusarium solani* (Ohta et *al.*, 1979). Compounds I–VI were isolated from culture filtrates. Other compounds are presumed catabolic intermediates.

lignin model compounds (e.g., ferulic acid, syringaldehyde, and vanillin) are biodegradable by mixed microbial cultures to methane and carbon dioxide under strict anaerobic conditions. These authors suggested that anaerobic degradation of lignin-related·aromatic compounds may be much more common than previously realized. However, it may be necessary that lignin be converted to low-molecular-weight phenols before it becomes accessible to anaerobic microorganisms (Hackett et *al.*, 1977).

6.3 INVOLVEMENT OF PHENOL OXIDASES AND QUINONE OXIDOREDUCTASES IN LIGNIN BIODEGRADATION

There are three distinct types of enzymes that are known as "phenol oxidases": laccase (O_2: *p*-diphenol oxidoreductase), peroxidase (donor: H_2O_2 oxidoreductase), and tyrosinase (O_2: *o*-diphenol oxidoreductase). Ander and Eriksson (1978) have recently reviewed the literature concerning these enzymes, and readers are referred to this excellent review for information about nomenclature and mechanisms of action of phenol oxidases. For this discussion it is sufficient to point out that (a) tyrosinases catalyze the monohydroxylation of phenols to yield o-diphenols or o-quinones (Figure 6.25A) or oxidize catechols forming o-quinones (Figure 6.25B), and (b) laccase and peroxidase catalyze the oxidation of both o- and p-diphenols by abstraction of an electron and a hydrogen ion from a hydroxyl group forming an aryloxy free radical (Figure 6.25C). The radicals formed can undergo various further transformations, including disproportionation or polymerization by radical coupling (Figure 6.25D).

There seems to be a correlation between lignin degradation and the ability of microorganisms to produce phenol oxidases (Ander and Eriksson, 1976). Numerous investigators have shown an almost universal presence of phenol oxidases in lignin-degrading fungi (Davidson et al., 1938; Sundman and Nase, 1971; Harkin and Obst, 1973). Even the "atypical" white-rot fungi such as Polyporus dichrous (=Gloeoporus dichrous) and Stereum frustalatum that were thought not to produce phenol oxidases are usually shown to, in reality, produce these enzymes (cf. Raiha and Sundman, 1975; Ander and Eriksson, 1978). Lignin-degrading actinomycetes have thus far not been shown to produce laccases (D.L. Crawford, personal communication), though some Streptomyces are known to produce tyrosinase (Suter et al., 1978). The lignin-degrading actinomycete Streptomyces viridosporus (Phelan et al., 1979) produces a dark soluble pigment when grown in the presence of

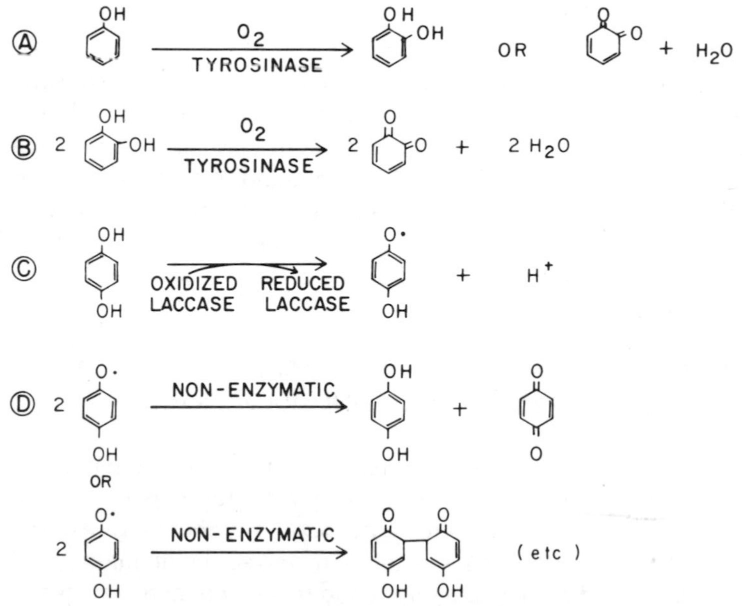

Figure 6.25 Activities of phenol oxidases (Ander and Eriksson, 1978).

L-tyrosine, and thus probably produces a tyrosinase (unpublished observation).

Kirk (1971) and Eriksson and Lindholm (1971) discussed the possible role of phenol oxidases in lignin degradation, concluding that phenol oxidases can mediate a certain amount of direct oxidative degradation of lignin, but that the primary effect of phenol oxidases is to further polymerize lignin and lignin degradation products (see Brunow et al., 1978; Kaplan, 1979). There is apparently no sound experimental evidence that phenol oxidases alone can mediate extensive lignin degradation (Gierer and Opara, 1973).

It is apparent from the work of Ander and Eriksson (1976) that phenol oxidases are in some manner involved in lignin degradation. These authors compared the lignin degradation abilities of a wild-type strain, a phenol oxidase–less mutant, and a phenol oxidase-positive revertant of Sporotrichum pulverulentum (=Phanerochaete chrysosporium) to determine whether phenol oxidase production is required for lignin degradation by white-rot fungi. The phenol oxidase-less mutant was unable to degrade lignin, while the revertant strain regained its ability to degrade lignin. Addition of laccase to kraft lignin agar plates also restored the lignin-degrading ability of the phenol oxidase-less mutant. Thus it appears that lignin degradation by Sporotrichum pulverulentum is dependent on the activity of a phenol oxidase.

The exact role played by phenol oxidases in lignin degradation remains to be firmly established. Ander and Eriksson (1976, 1978) have discussed the following possible roles: (a) phenol oxidases may play a role in detoxifying low-molecular-weight phenols released during lignin degradation (Gadd, 1957; Gierer and Opara, 1973; Grabbe et al., 1968; Law, 1959; and Rosch, 1966); (b) phenol oxidases may initiate lignin degradation by performing some critical initial chemical transformation, perhaps demethylation of methoxyl groups (Kirk and Chang, 1975; Nord, 1966; and Trojanowski and Leonowicz, 1969); (c) phenol oxidases may function in regulating the production of both lignin-degrading and polysaccharide-degrading enzymes (Ander and Eriksson, 1976); and (d) laccase and peroxidase may function in concert with cellobiose: quinone oxidoreductase ("cellobiose dehydrogenase") to regulate lignin biodegradation (Ander and Eriksson, 1976).

Shimada (1979) has pointed out that lignin is a racemic polymer; thus lignin-degrading microorganisms must have some enzymic device to circumvent the asymmetric structures of the optically inactive substructures of lignin. Shimada (1979) suggests that phenol oxidases may play a role in removing the original asymmetric centers in the lignin molecule by catalyzing the nonstereospecific oxidation of benzyl alcohol moieties to form benzyl ketones (Figure 6.26). Further degradation of the lignin polymer could then proceed by utilization of many fewer enzymes than would be required to degrade the stereospecifically complex natural polymer. These arguments are based on the assumption that lignin-degrading enzymes are likely to be highly

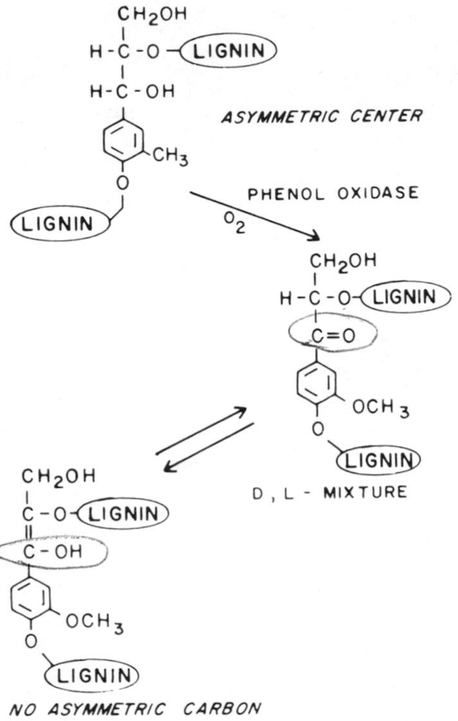

CH₂OH
H-C-O-(LIGNIN)
H-C-OH

ASYMMETRIC CENTER

CH₃

O

(LIGNIN)   O₂   PHENOL OXIDASE

CH₂OH
H-C-O-(LIGNIN)
(C=O)

OCH₃
O
(LIGNIN)

D , L - MIXTURE

CH₂OH
C-O-(LIGNIN)
(C-OH)

OCH₃
O
(LIGNIN)

NO ASYMMETRIC CARBON

Figure 6.26 A hypothetical role for phenol oxidases in the fungal degradation of lignin (Shimada, 1979).

stereospecific (the usual case for enzymic reactions; Bentley, 1970). Though Shimada (1979) presents some evidence to support his hypothesis, much more work is required before this hypothetical mechanism is established as fact.

Cellobiose dehydrogenase is a recently discovered enzyme that is involved in wood degradation (Westermark and Eriksson, 1974a,b). The enzyme appears to be of importance for the degradation of both cellulose and lignin (Ander and Eriksson, 1978). Cellobiose dehydrogenase is a flavoprotein with FAD as its prosthetic group. The enzyme simultaneously reduces a quinone during oxidation of cellobiose (Figure 6.27). The quinone requirement is somewhat nonspecific, with both o- and p-quinones serving as substrates. As Figure 6.27 illustrates, the concomitant oxidation of cellobiose and reduction of quinones by cellobiose dehydrogenase can be coupled to reoxidation of product phenols by laccase, producing a quinone ⟷ phenol oxidation/ reduction cycle. The result of such a coupling is oxidation of large quantities of cellobiose to cellobiono-δ-lactone, with phenols being used in catalytic amounts.

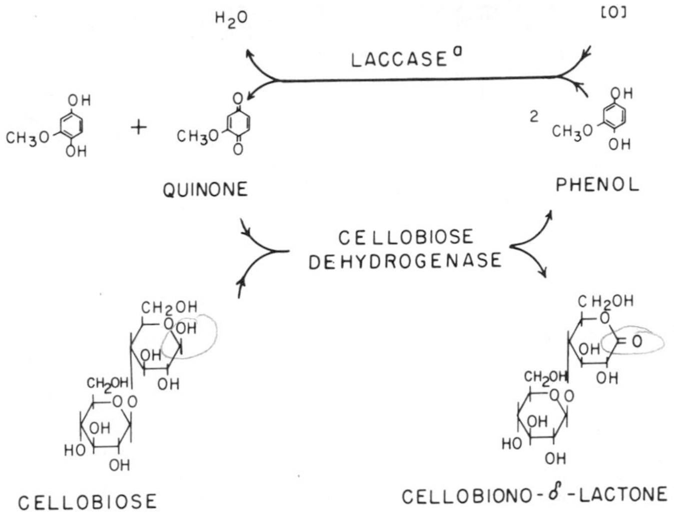

Figure 6.27   Reactions catalyzed by fungal cellobiose : quinone oxidoreductase and laccase (Ander and Eriksson, 1978). See Figure 6.25C and 6.25D.

Activity of white-rot fungal cellobiose dehydrogenase is dependent on degradation of cellulose as a source of cellobiose. Thus if cellobiose dehydrogenase is absolutely required for lignin degradation, this would explain why lignin degradation by white-rot fungi is always accompanied by some degradation of wood polysaccharides. This is apparently not the case. Ander and Eriksson (1975) have shown that a mutant of *Sporotrichum pulverulentum* (=*Phanerochaete chrysosporium*) lacking both cellulase and cellobiose dehydrogenase activities is still able to degrade kraft lignin and wood lignins, although more slowly than the wild-type fungus.

The role played by cellobiose dehydrogenase in lignin degradation is not clear. Eriksson and co-workers (Ander and Eriksson, 1978) have suggested that the primary role of this enzyme may be in the detoxification of lignin-derived quinones, which are known in some instances to strongly inhibit ring-cleaving enzymes (Bilton and Cain, 1968). This hypothesis is supported by the observations of Ander and Eriksson (1977), who found that 25 different white-rot fungi produce cellobiose dehydrogenase, while brown-rot fungi do not produce the enzyme at all (Ander and Eriksson, 1978). It has been suggested that a primary difference between white-rot and brown-rot fungi is that the latter group lacks the efficient system of ring-fission enzymes that is produced by the former group.

It is also conceivable the cellobiose dehydrogenase is involved in extracellular cycling of reducing power required for mixed-function oxidations of lignin (Figure 6.28). There have apparently been no published experiments to

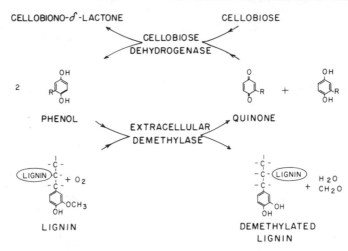

Figure 6.28   A hypothetical role for cellobiose dehydrogenases in lignin biodegradation.

test this possibility when the coupling activity observed for laccase might also be performed by other enzymes such as mixed-function oxygenases.

6.4   PHYSIOLOGICAL STUDIES OF LIGNIN DEGRADATION BY *PHANEROCHAETE CHRYSOSPORIUM*

Only one microorganism has been subjected to a thorough examination of the physiological controls governing lignin degradation. That organism, *Phanerochaete chrysosporium,* has been extensively studied by Kirk and his co-workers. In a number of elegant experiments using ¹⁴C-labeled lignins these investigators have been able to optimize lignin degradation by this white-rotter while learning much concerning its physiological control mechanisms for lignin biodegradation. It is worthwhile at this point to summarize these important studies.

Among the initial discoveries of Kirk and his colleagues (Kirk *et al.,* 1976) was that lignin cannot serve as a sufficient carbon and energy source for its own catabolism by *Phanerochaete chrysosporium.* The fungus only degrades lignin when supplied with an additional growth substrate such as cellulose, glucose, or succinate. Lignin degradation is in fact directly proportional to the amount of additional growth substrate provided. The physiological basis of this requirement is unclear; however, since lignin degradation by the *Phanerochaete* is apparently an event associated with "secondary metabolism" (metabolic activities that occur only after primary growth is complete;

see following), physiological control over the ligninolytic enzyme system may directly preclude primary growth on lignin (Kirk and Fenn, in press).

Agitation of cultures greatly suppresses lignin degradation by *Phanerochaete chrysosporium* (Kirk et al., 1978). This observation indicates that close contact between lignin particles and fungal hyphae is required for lignin degradation.

Lignin degradation by the *Phanerochaete* has a high requirement for oxygen. Conversion of ^{14}C-lignin to $^{14}CO_2$ does not occur in the presence of 5% O_2 in N_2 or at O_2 concentrations lower than 5%, but degradation is good at 21% O_2 (ambient O_2 concentration) (Kirk et al., 1978). Lignin degradation is enhanced two- to threefold under 100% O_2, but maximal degradation seems to occur at 40–60% O_2 concentrations (Kirk and Fenn, in press).

Lignin degradation by *Phanerochaete chrysosporium* is very pH dependent. Conversion of ^{14}C-DHP's to $^{14}CO_2$ occurs maximally between pH 4.0 and 4.5, though the pH optimum for growth is somewhat higher (Kirk et al., 1978). We have shown that the pH optimum for conversion of ^{14}C-[LIGNIN]-spruce to $^{14}CO_2$ by *Phanerochaete chrysosporium* is also about 4.5 (Figure 6.29).

Among the most interesting discoveries concerning lignin degradation by *Phanerochaete chrysosporium* is that conversion of ^{14}C-lignin to $^{14}CO_2$ is suppressed by nutrient nitrogen, if nitrogen concentrations exceed a subsistence level (about 2.4 mM) (Kirk et al., 1978). The source of nitrogen is usually

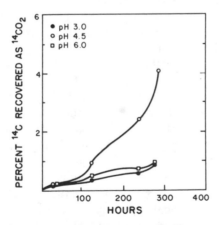

Figure 6.29 Degradation of ^{14}C-[LIGNIN]-spruce by *Phanerochaete chrysosporium* as a function of pH. Incubations were performed in duplicate in 100 ml of minimal medium containing 0.002% (w/v) thiamine and buffered with 0.01M aconitic acid. Each flask received approximately 20 mg (9749 dpm) of ^{14}C-[side chain, LIGNIN]-spruce (autoclaved dry) prepared as described by Robinson and Crawford (1978). Incubations were performed at 37°C. Uninoculated controls evolved no significant $^{14}CO_2$. Final pH's were unchanged from initial values.

not critical, with most utilizable nitrogen sources showing repression of the fungal ligninolytic enzyme system. Glutamate and NH_4^+ are particularly good suppressors of lignin decomposition. Repression of lignin degradation probably does not occur through a direct effect on the ligninolytic system itself, but is indirect by way of control of secondary metabolism (Kirk and Fenn, in press). Nitrogen starvation triggers secondary metabolism in *Phanerochaete chrysosporium;* and since lignin degradation is a secondary metabolic process, nitrogen starvation induces the ligninolytic system along with all other such secondary processes (e.g., *de novo* synthesis of veratryl alcohol; Lundquist and Kirk, 1978). Degradation of lignin's methoxyls, side chains, and rings is repressed similarly by nutrient nitrogen, indicating that the complete ligninolytic system is under the same control mechanism.

Kirk and Fenn (in press) have suggested that it may be of advantage to wood-rotting fungi to degrade lignin only during secondary metabolism, since wood contains very little nitrogen and primary growth of the fungus in wood would be very transient (occurring only until the small amount of available nitrogen is depleted).

The ligninolytic enzyme system is not induced by the presence of lignin within the growth medium of *Phanerochaete chrysosporium* (Keyser et al., 1978). Induction is solely initiated by nitrogen starvation and is independent of the carbon source being used for growth. Functioning of the ligninolytic system does require protein synthesis just prior to initiation of secondary metabolism, as shown by the use of cycloheximide to delay onset of lignin degradation even under conditions of nitrogen starvation.

Recently Reid (1979) has shown that degradation of ^{14}C-[LIGNIN]-aspen wood to $^{14}CO_2$ by *Phanerochaete chrysosporium* is also suppressed by excess nutrient nitrogen, confirming the ^{14}C-DHP work of Kirk and colleagues. Reid suggested that the carbon–nitrogen ratio of the growth medium was more important as an influence to lignin degradation rates than the absolute carbohydrate and nitrogen levels. He also found that sulfate and phosphate limitation of growth did not stimulate lignin catabolism.

Kaplan and Hartenstein (1980) examined degradation of specifically labeled ^{14}C-DHP's by seven white-rot fungi. They reported rather low conversions of ^{14}C-DHP to $^{14}CO_2$ after 2 months of incubation (<1.0% to a maximum of 10.82%). These results were probably caused by nitrogen repression of the various fungal ligninolytic enzyme systems, as the growth medium employed contained 2.0 g/l of $NH_4H_2PO_4$ (17 mM) and 1.0 g/l of yeast extract (organic nitrogen) with only 100 mg of cosubstrate cellulose provided. The cultures were probably never nitrogen limited.

It is also known that (a) o-phthalate (sometimes used as a media buffering agent) inhibits lignin degradation by *Phanerochaete chrysosporium* at least 50% at 0.01M (Fenn and Kirk, 1979; 2,2-dimethylsuccinate is suggested as a substitute buffer for o-phthalate), (b) the level of other media nutrients such as phosphate does not affect the appearance or level of the *Phanerochaete*'s ligninolytic system (Kirk and Fenn, in press), and (c) lignin carbon enters the central metabolic carbon pool of *Phanerochaete chrysosporium* (Kirk and Fenn, in press).

6.5 SPECULATION CONCERNING THE PATHWAYS FOR MICROBIAL LIGNIN DEGRADATION

There are two generally discussed mechanisms whereby microorganisms might degrade lignin: (a) depolymerization of the lignin macromolecule with release of monomeric and dimeric lignin fragments which are transported into

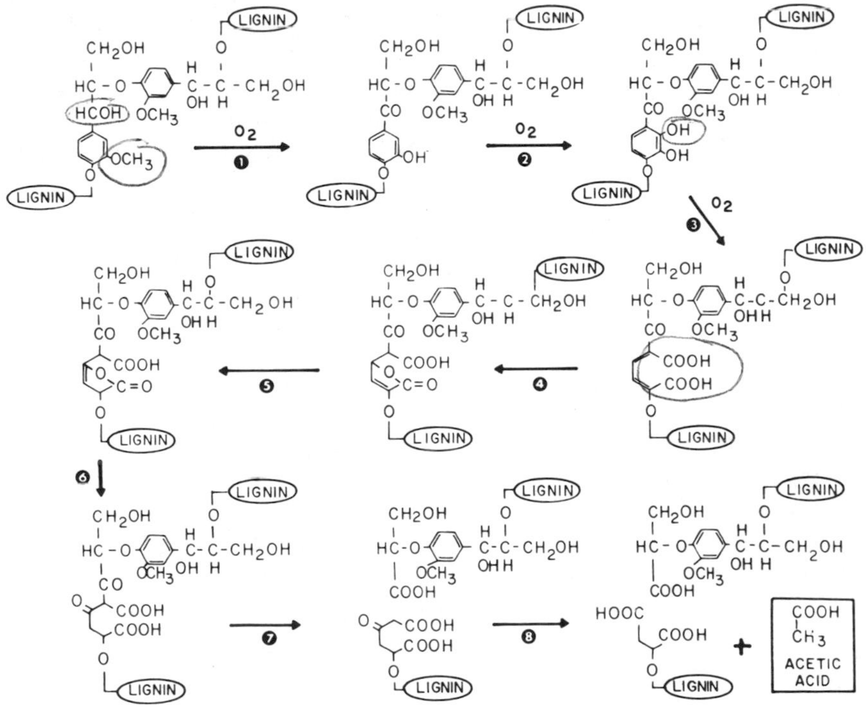

Figure 6.30 A hypothetical catabolic sequence for the microbial degradation of lignin. (1) Demethylation of lignin's 3-0-methyl groups and oxidation of benzyl alcohols to ketones; (2) aromatic hydroxylation to form catechol structures; (3) aromatic ring fission by orthofission; (4) ring lactonization (see structure II, Figure 4.1); (5–8) reactions directly analogous to those of the β-ketoadipate pathway (Figure 6.1), ultimately yielding simple organic acids such as acetate. This erosion of the lignin molecule would occur at all lignin surfaces available to the microorganism involved. This scheme necessitates the involvement of extracellular or cell surface–bound enzymes (see Section 4.1.1).

microbial cells where they are degraded, and (b) dearomatization of the intact polymer by cleavage of rings while they are still bound in the macromolecule, followed by the erosion of the resulting polymeric, aliphatic network. The meager evidence accumulated thus far is supportive of the latter alternative.

It is important to remember that only a few organisms (notably, *Phanerochaete chrysosporium*) have been examined in any detail as regards the pathways of lignin biodegradation. Because catabolic variations between superficially similar microorganisms are often immense, all organisms should not be expected to utilize similar degradative mechanisms. It is not unlikely that lignin degradation may proceed in nature by more than one mechanism.

Details of even the best understood lignin decay mechanism are almost nonexistent. For example, knowledge of the enzymology of lignin degradation is, to be generous, rudimentary. Even the role of the massively studied phenol oxidases in lignin degradation is still open to serious debate. Most available evidence concerning lignin degradation pathways is either circumstantial or indirect. With these limitations taken into consideration, a reasonable, but very hypothetical catabolic sequence for microbial lignin degradation is presented in Figure 6.30.

Chapter 7

THE FUTURE FOR LIGNIN BIODEGRADATION RESEARCH

Efforts to arrive at a better understanding of the mechanisms whereby microorganisms degrade lignin have now been underway for more than 80 years. Progress toward this understanding, however, has been agonizingly slow. Fortunately, this undesirable situation recently has been changing for the better. In recent years several serious impediments to progress in lignin biodegradation research have been largely overcome. Chief among these advances have been (a) development of a relatively complete understanding of the chemical structures of various lignins, and (b) development of unequivocal assays for quantifying rates and extents of lignin mineralization to CO_2. With these advances came corresponding strides in our understanding of lignin decay mechanisms. In fact, progress has been so rapid that even the frequent appearance in the literature of review papers concerning lignin biodegradation has failed to keep pace. Virtually every review has been seriously out of date either before or shortly after its appearance. We can only hope that such progress continues unabated.

Despite this optimistic appraisal of the present status of lignin biodegradation research, there are still some serious problems to be overcome. Some of these problems are of such a serious nature that they threaten to dampen the present rapid progression in our knowledge of lignin decay processes. In the following section I point out some of the problems that I feel are important enough to warrant concerted research efforts during coming years.

7.1 ENZYMOLOGY OF LIGNIN BIODEGRADATION: ENZYME ASSAY TECHNIQUES

One area of lignin biodegradation research that has not seen rapid progress concerns the enzymological mechanisms of lignin decay. Most of our evidence concerning lignin-transforming enzymes (Chapter 6) is indirect (e.g., studies of degradation of lignin model compounds and chemical characterizations of decayed lignins). We have yet to actually isolate and characterize any protein that has been shown unequivocally to be a component of a microbial ligninolytic enzyme system.

We do know enough about the chemistry of lignin degradation to predict with some certainty the types of enzymes involved in lignin biodegradation. As Chapter 6 indicates, these ligninolytic enzymes probably include dioxygenases, monooxygenases, dehydrogenases, aldolases, and similar proteins. Why then has it been so difficult to actually demonstrate such enzymatic activities in enzyme preparations prepared from ligninolytic microorganisms? Our failure in this regard may reflect our inability at present to overcome some major technical problems hindering development of enzymatic assays.

For example, the following questions repeatedly come up as we attempt to devise assays for ligninolytic monooxygenases. Do such enzymes require cofactors? If so, what are those cofactors? All the classical catabolic monooxygenases studied thus far (e.g., aromatic hydroxylases; R. L. Crawford, 1976a) and aromatic demethylases (Ribbons, 1970) require cofactors as a source of reducing power. These cofactors are usually reduced pyridine nucleotides [NAD(P)H] that function in concert with soluble (cytoplasmic) monooxygenases. It is unlikely that reduced pyridine nucleotides function as cofactors with ligninolytic monooxygenases if, as most investigators feel, these enzymes are extracellular. This would require excretion of NAD(P)H into the growth medium, not an energetically reasonable process. It may be, however, that ligninolytic monooxygenases are not actually extracellular, but are cell surface bound. This might allow for the functioning of unusual cell wall bound pyridine nucleotides. Alternatively, ligninolytic monooxygenases might contain electron donors/acceptors as prosthetic groups (FAD?) within the enzymes' tertiary structure, thus negating the problem of excretion of cofactors into the surrounding environment. Other explanations for presumed cofactor requirements (e.g., quinone-hydroquinone interconversions, Figure 6.28) are of course also possible. Finally, ligninolytic monooxygenases may not be classic mixed-function oxygenases at all, having their own unique mechanism that does not require exogenous reducing power (perhaps the oxidative catalysts are not even enzymes?). We simply do not know the answers to such questions. These questions, however, must be resolved before we can hope to develop actual assay techniques to detect activities of the demethylases and hydroxylases that are assumed to be part of microbial ligninolytic enzyme systems.

7.2   ENZYMOLOGY OF LIGNIN BIODEGRADATION: LOW ACTIVITIES OF LIGNINOLYTIC ENZYMES

Another serious problem associated with detection of ligninolytic enzymes concerns the apparent low activities of these enzymes within lignin-degrading microorganisms. For example, even the most active ligninolytic cultures of *Phanerochaete chrysosporium* degrade only a few micrograms of ^{14}C-DHP to $^{14}CO_2$ per day. Thus measurement of individual components of the ligninolytic system (between ^{14}C-DHP and $^{14}CO_2$) requires extremely sensitive assay procedures. Use of typical assay techniques [e.g., the use of an oxygen electrode to measure dioxygenase activities (R. L. Crawford et al., 1979a) in crude cell-free extracts of bacteria] simply do not allow sufficient sensitivity. Therefore present assay procedures must be improved to bring them up to the required sensitivity (Weinstein and Gold, 1979). This is a difficult task that will require some ingenuity. Some possible solutions to sensitivity problems lie in the use of radioisotopically labeled lignin model compounds. The models chosen as substrates must of course be relevant to (degraded by) the ligninolytic system of the organism under study [e.g., ^{14}C-veratrylglycerol-β-guaiacyl ether is probably degraded by the ligninolytic system of *Phanerochaete chrysosporium* (Weinstein et al., 1980)] These models conceivably could be used in sensitive radiotracer trapping experiments, where labeled enzyme reaction products (e.g., demethylated veratrylglycerol-β-guaiacyl ether) are trapped by addition of unlabeled carrier (e.g., guaiacylglycerol-β-guaiacyl ether) which is then reisolated and any trapped radioactivity quantified. Such assays should be adaptable to both whole-cell and cell-free extract based experiments, as well as to use in detection of extracellular enzyme activities.

7.3   ENZYMOLOGY OF LIGNIN BIODEGRADATION: PREPARATION OF ACTIVE CELL EXTRACTS

Another possible explanation for our inability to produce cell-free extracts containing active ligninolytic enzymes is that these enzymes are intrinsically unstable. If ligninolytic enzymes are structurally incorporated into the cell walls or membranes of lignin-degrading microorganisms, then any form of mechanical disruption (e.g., by sonication or grinding of mycelial pellets) of the enzyme "complexes" may irreversibly disaggregate them and thereby render them inactive. Thus more research needs to be pursued in developing mild lysis procedures for basidiomycetes such as *Phanerochaete chryso-*

sporium (see Jefferies *et al.*, 1977), so that active enzyme complexes, if these are required, may be released without inactivation.

7.4   BIOCHEMSTIRY OF LIGNIN BIODEGRADATION: OTHER PROBLEMS AND TECHNIQUES

7.4.1   Other Problems

There are several other nagging questions that deserve some particular emphasis in future research efforts. Some that come to mind are: (a) What is the actual role, if any, played by phenol oxidases in the processes of lignin degradation? (b) What is the actual role of cellobiose-quinone oxidoreductase in lignin degradation? (c) Which lignin model compounds are actually relevant to the ligninolytic enzyme system of specific lignin-degrading microorganisms? (d) Is the study of kraft lignin degradation relevant to the study of degradation of natural lignins?

7.4.2   Use of Genetic Techniques

The use of genetic techniques in the study of lignin biodegradation is an area where large strides in our understanding of lignin biodegradation are possible. Such techniques have hardly been applied to the problem, even though they hold great potential for answering some of our most difficult questions. For example, the use of so-called catabolically blocked mutants of lignin-degrading microorganisms could potentially lead to a complete mapping of a ligninolytic enzyme system. In this technique microbial mutants would be produced that have individual ligninolytic enzymes inactivated by a mutational event. Such organisms would then oxidize lignin up to the blocked enzyme, accumulating a lignin enriched in identifiable chemical structures that normally would be substrates for the inactivated protein. In theory mutants blocked at every enzyme in the ligninolytic pathway could be isolated. Chemical characterization of lignins modified by all the mutants would then allow postulation of a complete catabolic scheme. These techniqes could also be applied to produce blocked mutants for degradation of relevant lignin model compounds. If the models are truly relevant to the ligninolytic system, then mutants blocked for model compound degradation steps (e.g., demethylation) also should be blocked in similar steps of the ligninolytic system. Blocked mutant procedures have served admirably in studies of microbial intermediary metabolism. They offer similar hopes of progress in elucidating the mechanisms of lignin biodegradation. Future lignin biodegradation research should therefore emphasize these heretofore little-used procedures. A few investigators have already realized the great potential that lies in the use of genetic techniques (Ander and Eriksson, 1976; Gold and Cheng, 1978). More such investigators are needed.

7.5 APPLICATIONS OF LIGNIN-DEGRADING MICROORGANISMS TO INDUSTRIAL-SCALE BIOCONVERSION OF LIGNOCELLULOSE

Despite much effort toward the goal of developing large-scale applications of lignin-degrading microorganisms to bioconversion of lignocelluloses to other more valuable commodities, no real success story can be reported at this time. The potential benefits from such microbial applications are so great, however, that in any list of future research problems, development of industrial-scale bioconversion processes must be included. In fact, this goal should be envisioned as a long-term commitment, rather than as a short-term goal.

7.5.1 Gasohol from Lignocellulosic Wastes

There are numerous examples of desirable bioconversions of lignocellulosics (Chapter 1) to useful products. However, for this discussion one particularly important example is emphasized, the conversion of biomass to ethanol. This example emphasizes (a) the great benefits that are possible from bioconversion of lignocellulose, (b) the problems that arise in attempts at bioconversion, and (c) the necessity for long-term commitments to bioconversion research.

The production of ethanol from biomass for use as a fuel (e.g., in gasohol) has a great and highly recognized potential for alleviating some of the present and anticipated strains on this country's liquid fuels supply. Presently, most proponents of ethanol from biomass envision using food crops such as grains, and potatoes as raw feed stocks for ethanol production. Though this approach may be of short-term benefit, it may not be so wise over the long term. Even when problems of net energy balance (energy output minus energy input to the process; Hopkinson and Day, 1980) are not considered, it is probably unwise to tie our energy needs into our food chain by using food crops as raw materials. We might conceivably become so dependent on agriculture to furnish a significant fraction of our liquid fuel supply that increases in worldwide demand for food might simultaneously disrupt both food and energy distribution systems (with devastating consequences). A much safer and readily available feedstock for ethanol production is waste lignocellulose (Table 1.2; Sitton et al., 1979).

Unfortunately, the technologies for using lignocellulose as a feedstock for ethanol production have not been sufficiently developed to make ethanol production from lignocellulose economically viable. Lignocelluloses are very resistant to fermentative bioconversion to ethanol. This is because lignin itself is not readily degradable under anaerobic conditions, and it acts as a barrier blocking the efficient fermentation of cellulose, the principal precursor of ethanol from lignocellulose. Currently the only practical way to utilize lignocellulose to produce ethanol is to (a) chemically or mechanically delignify the lignocellulose (with loss of up to 30–50% of the feedstock), (b) hydrolyze the delignified cellulose to the level of free sugars, and (c) ferment the free sugars to ethanol. The sum of these processes is too costly, poses pollution problems (waste disposal), and is an inefficient use of the original feedstock (most of the lignin and some of the cellulose is lost in the first step). We must find better bioconversion methods (e.g., Sitton et al., 1979, appear to have an improved method for production of ethanol from agricultural lignocellulose; Figure 7.1).

We need to discover ways to more efficiently use both the cellulose and lignin in lignocellulosics, particularly waste lignocelluloses. For example, it would be a very significant breakthrough if someone developed an economically viable process that produced some valuable chemical or commodity from the lignin of wood wastes while the residual cellulose was converted to ethanol. One approach to achieving such a success is to use a series of microorganisms to bioconvert waste lignocellulose. For example, a white-rot fungus might be used to delignify wood wastes (producing single cell protein) which could then be hydrolyzed enzymatically or chemically to produce a glucose solution which in turn would be fermented by a yeast to produce ethanol and more single cell protein. As Chapter 1 points out, much research has already been performed on this and similar bioconversion sequences. Yet no such bioconversion process has become an economic success.

Lack of success is probably not a result of faulty process design. Microorganisms are available to perform all the necessary bioconversions. For example, enzymatic cellulose hydrolysis procedures and yeast fermentation processes are becoming highly developed and efficient. The first step in the bioconversion sequence (biological delignification), however, is not so well understood (Chapter 6), and this step is critical to the next (enzymatic hydrolysis) step. Research has shown that there are numerous organisms that will selectively delignify wood. However, the biological delignification process is generally too slow to permit development of a successful industrial procedure. We thus need to speed up the delignifying capabilities of selected ligninolytic microorganisms. To accomplish this goal requires a better understanding of (a) the physiology of ligninolytic microorganisms, (b) the enzymology of lignin biodegradation, and (c) the mechanisms of genetic control over ligninolytic enzyme systems. To attain this basic understanding requires a long-term commitment to scientific study of lignin biodegradation processes. Once we attain a fundamental understanding of lignin biodegradation processes and then apply our knowledge to improving rates of biological delignification, industrial production of ethanol and other products from wood wastes will probably become a reality.

OPERATING COSTS FOR 4.5 M gal/yr ETHANOL PROCESS	
	$ / YR
Cornstalks ($15/ton)	1,095,000
Acid recovery	300,700
Make up acid	421,700
Utilities	381,200
Neutralizer	265,000
Yeast extract and nutrients	256,000
Labor	409,400
Maintenance	409,400
Depreciation, taxes, insc.	818,800
	4,357,200
Production (gal/yr)	4,510,000
Break even price ($/gal)	$ 0.966
R.O.I. at $1.185/gal	15.25%

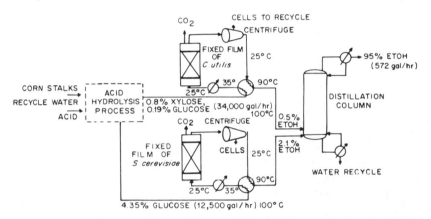

Figure 7.1 Proposed process for production of ethanol from agricultural residues (Sitton et al., 1979). Cost estimates are for 1978–1979. As the prices of liquid fuels escalate, cost effectiveness of bioconversion processes such as this will increase. Capital outlay costs are not included here, but the authors estimate that capital investment for a 4.5 $\overline{M}$ gal/yr ethanol process might be about 7.53 million dollars.

The same reasoning applies to any other proposed bioconversion process where delignification is a critical step (e.g., bacterial single cell protein production, methane fermentations, and production of forage for ruminants). We must attain a better understanding of how microorganisms degrade lignin before we can hope to realize a successful industrial application of bioconversion. Thus I cannot overstate the case for a long-term, sustained effort at understanding at the most fundamental level how microorganisms decompose this complex material we call lignin.

7.5.2 Other Bioconversions of Lignocellulose

A major problem that has hindered bioconversion research has been a shortage of imaginative ideas for bioconversion processes and goals. Much effort has been rightfully expended in trying to convert lignocellulosics to obvious products like ethanol, methane, and single-cell protein. Much less effort has been applied to the study of more novel chemicals and processes. The following section illustrates an example (that may or may not be feasible) of a novel bioconversion process that embodies this idea of trying something new.

It is conceivable that the polysaccharidic components of waste lignocellulosics might be bioconvertable to chemical feedstocks for production of biodegradable plastics. The first step in such a process would be the production of a sugar solution from a lignocellulose (e.g., by the acid hydrolysis method of Sitton et al., 1979). A microorganism that produces the polyester poly-β-hydroxybutyric acid as storage material would then be grown on the sugar solution. An optimal organism would be similar to *Azotobacter beijerinckii* NCIB 9067, which produces poly-β-hydroxybutyric acid in excess of 70% of its dry weight when it is grown on glucose and then exposed to conditions of O_2 limitation (Dawes and Senior, 1973). This organism also fixes atmospheric N_2 and thus requires no expensive combined nitrogen sources in its growth medium. An industrially attractive continuous fermentation process with such an organism would appear promising. A two-stage chemostatic fermentor could be used. The organism would be maintained in log phase in the first stage (Dawes and Senior, 1973), which would then feed into a second fermentor where O_2 limitation of growth would induce synthesis of high levels of poly-β-hydroxybutyric acid $[(-CH(CH_3)CH_2C(O)O-)_n]$. The outflow from stage two would optimally contain bacterial cells ($\sim 10^9$/ml) of $> 70\%$ polyester content. These cells could be harvested and the polyester component used as a chemical feedstock.

As Braunegg et al. (1978) point out, microbial poly-β-hydroxybutyric acid has a high potential for use as a biodegradable plastic substitute. In fact, Baptist et al. (1962) and Baptist and Werber (1963, 1965) have been issued patents for processing microbial poly-β-hydroxybutyric acid into molded plastic materials. These patents include methods for purification of microbial poly-β-hydroxybutyric acid and production of brittle and/or flexible (plasticized) plastic substitutes. In one of these patents (Baptist and Werber, 1963) the authors report that poly-β-hydroxybutyric acid containing bacteria (40% or higher of polyester content) can be molded directly into useful plastic substitutes without prior purification of the polyester.

Thus it appears technically feasible to produce chemical feedstocks for the plastics industry by bioconversion of plentiful waste lignocellulosics. Alternatively, microbial poly-β-hydroxybutyric acid prepared from waste lignocellulose could be converted to other hydrocarbon-based chemical feedstocks such as β-hydroxybutyric acid (by hydrolysis of the polyester) or crotonic acid (by distillation of β-hydroxybutyric acid, which dehydrates during distillation). All these processes are already technically feasible since the majority of the methodologies have been worked out by various independent groups (Baptist, 1962; Baptist and Werber, 1963, 1965; Dawes and Senior, 1973; Braunegg et al., 1978; and Sitton et al., 1979).

The main point in the above discussion is not whether this particular suggestion is a good one, but that unusual bioconversion possibilities ought to be considered more often. We have much lignocellulose available for bioconversion to useful products. Much of the technical background for doing so has already been accumulated, though in numerous and scattered places. We need more imaginative people to bring these scattered technologies together.

REFERENCES

Adams, G. A., and G. A. Ledingham. 1942. Biological decomposition of chemical lignin. I. sulfite waste liquor. *Can. J. Res. Sect. C* **20**: 1–12.

Adler, E. 1968. Liginets kemiska byggnad. *Kem. Tidskr.* **80**: 279–290.

Adler, E. 1977. Lignin chemistry—past, present and future. *Wood Sci. Technol.* **11**: 169–218.

Alexander, M. 1965. Most probable number technique. In *Methods of Soil Analysis, Part 2*. C. A. Black, ed. American Society of Agronomy, Madison, Wisc., pp. 1467–1472.

American Society for Testing and Materials. 1962. *Annual Book of ASTM Standards, Part 6*. Standard method for accelerated laboratory test of natural decay resistance of woods. Designation D 2017-63.

Ander, P., and K.-E. Eriksson. 1971. Selective degradation of wood components by white-rot fungi. *Physiol. Plant.* **41**: 239–248.

Ander, P., and K.-E. Eriksson. 1975a. Mekanisk massa Från Förrötad flisen inledande undersökning. *Sven. Papperstid.* **78**: 641–642.

Ander, P., and K.-E. Eriksson. 1975b. Influence of carbohydrates on lignin degradation by the white-rot fungus *Sporotrichum pulverlentum*. *Sven. Papperstid.* **78**: 643–562.

Ander, P., and K.-E. Eriksson 1976. The importance of phenol oxidase activity in lignin degradation by the white-rot fungus *Sporotrichum pulverulentum*. *Arch. Microbiol.* **109**: 1–8.

Ander, P., and K.-E. Eriksson. 1977. Selective degradation of wood components by white-rot fungi. *Physiol. Plant.* **41**: 239–248.

Ander, P., and K.-E. Eriksson. 1978. Lignin degradation and utilization by micro-organisms. *Prog. Ind. Microbiol.* **14**: 1–58.

Ander, P., A. Hatakka, and K.-E. Eriksson. 1980. Vanillic acid metabolism by the white-rot fungus *Sporotrichum pulverulentum*. *Arch. Microbiol.* **125**: 189–202.

Apenitis, A., H. Erdtman, and B. Leopold. 1951. Studies on lignin. V. The decay of spruce wood by brown-rotting fungi. *Sven Kem. Tidskr.* **63**: 195–207.

Ban, S., and M. Glanser-Soljan. 1979. Rapid biodegradation of calcium lignosulfonate by means of a mixed culture of microorganisms. *Biotechnol. Bioeng.* **21**: 1917–1928.

Baptist, J. N. 1962. U.S. Patent 3,044,942.

Baptist, J. N., and F. X. Werber. 1963. U.S. Patent 3,107,172.

Baptist, J. N., and F. X. Werber. 1965. U.S. Patent 3,182,036.

Barnes, C., and J. Friend. 1975. The lack of movement of [^{14}C]-phenylalanine and [^{14}C]-cinnamate after administration to leaves of *Polygonium* and wheat. *Phytochemistry* **14**: 139–142.

Bayly, R. C., S. Dagley, and D. T. Gibson, 1966. The metabolism of cresols by species of *Pseudomonas*. *Biochem. J.* **101**: 293–301.

119

Beal, F. C., W. Merrill, R. C. Baldwin, and J.-H. Wang. 1976. Thermogravimetric evaluation of fungal degradation of wood. *Wood Fiber.* **8:** 159–167.

Bellamy, W. D. 1972a. *Chem. Eng. News.* **50:** 14.

Bellamy, W. D. 1972b. The use of thermophilic microorganisms for the recycling of cellulosic waste. *Proc. Am. Inst. Chem. Eng.,* Meeting, Aug. 28, 1972.

Bellamy, W. D. 1974. Single cell protein from cellulosic wastes. *Biotechnol. Bioeng.* **16:** 869–880.

Bentley, R. 1970. In *Asymmetry in Biology,* Vol. 2. Academic Press, New York.

Bilton, R. F., and R. B. Cain. 1968. The metabolism of aromatic acids by micro-organisms. A reassessment of the role of *o*-benzoquinone as a product of protocatechuate metabolism by fungi. *Biochem. J.* **108:** 829–832.

Björkman, A. 1956. Studies on finely divided wood. 1. Extraction of lignin with neutral solvents. *Sven. Papperstid.* **59:** 477–485.

Björkman, A. 1957a. Studies on finely divided wood. 3. Extraction of lignin-carbohydrate complexes with neutral solvents. *Sven. Papperstid.* **60:** 243–251.

Björkman, A. 1957b. Studies on finely divided wood. 5. The effect of milling. *Sven. Papperstid.* **60:** 329–335.

Björkman, A., and B. Person. 1957a. Studies on finely divided wood. 2. The properties of lignins extracted with neutral solvents from softwoods and hardwoods. *Sven. Papperstid.* **60:** 158–169.

Björkman, A., and B. Person. 1957b. Studies on finely divided wood. 4. Some reactions of the lignin extracted by neutral solvents from *Picea abies. Sven. Papperstid.* **60:** 285–292.

Blakley, E. R. 1974. The microbial degradation of cyclohexanecarboxylic acid: a pathway involving aromatization to form *p*-hydroxybenzoic acid. *Can. J. Microbiol.* **20:** 1297–1306.

Blanchette, R. A., and C. G. Shaw. 1978a. Associations among bacteria, yeasts, and basidiomycetes during wood decay. *Phytopathology.* **68:** 631–637.

Blanchette, R. A., and C. G. Shaw. 1978b. Management of forest residues for rapid decay. *Can. J. Bot.* **56:** 2904–2909.

Bocks, S. M. 1967a. Fungal metabolism. III. The hydroxylation of anisole, phenoxyacetic acid, phenylacetic acid and benzoic acid by *Aspergillus niger. Phytochemistry.* **6:** 785–789.

Bocks, S. M. 1967b. Fungal metabolism. I. The transformations of courmarin, *o*-courmaric acid and *trans*-cinnamic acid by *Aspergillus niger. Phytochemistry.* **6:** 127–130.

Boruff, C. S., and A. M. Buswell. 1936. The anaerobic fermentation of lignin. *J. Am. Chem. Soc.* **56:** 886–888.

Brandt, D., L. Hartz, and M. Mandels. 1972. Engineering aspects of the enzymatic conversion of waste cellulose to glucose. *Proc. Am. Chem. Eng.,* Meeting, Aug. 28, 1972.

Braunegg, G., B. Sonnleitner, and R. M. Lafferty. 1978. A rapid gas chromatographic method for the determination of poly-β-hydroxybutyric acid in microbial biomass. *Eur. J. Appl. Microbiol. Biotechnol.* **6:** 29–37.

Brauns, F. E. 1962. Soluble native lignin, milled wood lignin, synthetic lignin, and the structure of lignin. *Holzforschung.* **16:** 97–102.

Bray, M. W., and T. M. Andrews. 1924. Chemical changes of groundwood during decay. *Ind. Eng. Chem.* **16:** 137–139.

Brown, W., E. B. Cowling, and W. I. Falkehag. 1968. Molecular size distributions of lignins liberated enzymatically from wood. *Sven. Papperstid.* **22:** 811–821.

Brown, A. S., and A. C. Neish. 1955. Shikimic acid as a precursor in lignin biosynthesis. *Nature.* **175:** 688–689.

Brown, S. A., and A. C. Neish. 1955. Studies of lignin biosynthesis using isotopic carbon. IV. Formation from aromatic monomers. *Can. J. Biochem. Physiol* **33:** 948–962.

Brunow, G., H. Wallin, and V. Sundman 1978. A comparison of the effects of a white-rot fungus and H_2O_2-horseradish peroxidase on a lignosulfonate. *Holzforshung*. **32:** 189–192.

Buswell, J. A., P. Ander, B. Pettersson, and K.-E. Eriksson. 1979. Oxidative decarboxylation of vanillic acid by *Sporotrichum pulverulentum*. *FEBS Lett.* **103:** 98–101.

Buswell, J. A. and K.-E. Eriksson. 1979. Aromatic ring cleavage by the white-rot fungus *Sporotrichum pulverulentum*. *FEBS Lett.* **104:** 258–260.

Buswell, J. A., S. Hamp, and K.-E. Eriksson. 1979. Intracellular quinone reduction in *Sporotrichum pulverulentum* by a NAD(P)H: quinone oxidoreductase. *FEBS Lett.* **108:** 229–232.

Buswell, J. A., and A. Mahmood. 1972. Bacterial degradation of p-methoxybenzoic acid. *Arch. Mikrobiol.* **84:** 275–286.

Butler, J. H. A., and J. C. Buckerfield. 1979. Digestion of lignin by termites. *Soil Biol. Biochem.* **11:** 507–513.

Cartwright, N. J., and J. A. Buswell. 1967. The separation of vanillate o-demethylase from protocatechuate 3,4-oxygenase by ultracentrifugation. *Biochem. J.* **105:** 767–770.

Cartwright, N. J., and K. S. Holdom. 1973. Enzymic lignin, its release and utilization by bacteria. *Microbios.* **8:** 7–14.

Cartwright, N. J., and A. R. W. Smith. 1967. Bacterial attack on phenolic ethers. *Biochem. J.* **102:** 826–841.

Cerniglia, C. A., R. L. Hebert, P. J. Szaniszlo, and D. T. Gibson. 1978. Fungal transformation of naphthalene. *Arch. Microbiol.* **117:** 135–143.

Chahal, D. S., M. Moo-Young, and G. S. Dhillon. 1979. Bioconversion of wheat straw and wheat straw components into single-cell protein. *Can. J. Microbiol.* **25:** 793–797.

Chandra, S., and M. G. Jackson. 1971. A study of various chemical treatments to remove lignin from coarse roughages and increase their digestibility. *J. Agric. Sci., Camb.* **77:** 11–17.

Chang, H-M., E. Cowling, W. Brown, E. Adler, and G. Miksche. 1975. Comparative studies on celluloytic enzyme lignin and milled wood lignin of sweetgum and spruce. *Holzforschung*. **29:** 153–159.

Chang, H-M., and K. V. Sarkanen. 1973. Species variation in lignin: effect of species on rate of kraft delignification. *Tappi.* **56:** 132–134.

Chapman, P. J. 1972. An outline of reaction sequences used for the bacterial degradation of phenolic compounds, in *Degradation of Synthetic Organic Molecules in the Biosphere*, National Academy of Sciences, U.S. Government Printing Office.

Chapman, P. J. 1977. Bacterial degradation of methoxylated compounds via 4-methoxygentisic acid. *Abst. Ann. Meet., Am. Soc. Microbiol.* Q90, 276.

Chapman, P. J., and S. Dagley. 1962. Oxidation of homogentisic acid by cell-free extracts of a vibrio. *J. Gen. Microbiol.* **28:** 251–256.

Chattopadahyay, N. C., and B. Nandi. 1977. Degradation of cellulose and lignin in malformed mango inflorencence by *Fusarium moniliforme* var. *subglutinans*. *Wr. Rg. Acta Phytopath. Acad. Sci. Hung.* **12:** 283–287.

Chen, C. L., H-M. Chang, and T. K. Kirk. 1979. Lignin degradation products from spruce wood decayed by *Phanerochaete chrysosporium*, presented at a Symposium on Biosynthesis and Biodegradation of Cell Wall Components, ACS/CSJ Chemical Congress, April 1–6, Honolulu, Hawaii.

Chenna Reddy, C., and C. S. Vaidyanathan. 1974. Purification, properties and induction of a specific benzoate-4-hydroxylase from *Aspergillus niger* (VBC 814). *Biochem. Biophys. Acta.* **384:** 46–57.

Clausen, E. C., O. C. Sitton, and J. L. Gaddy. 1977. Bioconversion of crop materials to methane. *Process Biochem.* **12:** 5–8.

Commoner, B. 1979. The economics of methane. *Environment* **21:** 29–33.

Corbett, N. H. 1965. Micro-morphological studies on the degradation of lignified cell walls by ascomycetes and Fungi Imperfecti. *J. Inst. Wood Sci.* **14:** 18–29.

Cowling, E. B., 1961. Comparative biochemistry of decay of sweetgum sapwood by white-rot and brown-rot fungi. USDA Technical Bulletin 1258.

Cowling, E. B., and T. K. Kirk. 1976. Properties of cellulose and lignocellulosic materials as substrates for enzymatic conversion processes. *Biotechnol. Bioeng.* **6:** (Symposium M): 95–123.

Crawford, D. L. 1978. Lignocellulose decomposition by selected *Streptomyces* strains. *Appl. Environ. Microbiol.* **35:** 1041–1045.

Crawford, D. L., and R. L. Crawford. 1976. Microbial degradation of lignocellulose: the lignin component. *Appl. Environ. Microbiol.* **31:** 714–717.

Crawford, D. L., and R. L. Crawford. 1980. Microbial degradation of lignin. *Enzol. Microb. Technol.* **2:** 11–21.

Crawford, D. L., R. L. Crawford, and A. L. Pometto III. 1977a. Preparation of specifically labeled ^{14}C-[Lignin]- and ^{14}C-[Glugan]-lignocelluloseses and their decomposition by the microflora of soil. *Appl. Environ. Microbiol.* **33:** 1247–1251.

Crawford, D. L., S. Floyd, A. L. Pometto, III, and R. L. Crawford. 1977b. Degradation of natural and kraft lignins by the microflora of soil and water. *Can J. Microbiol.* **23:** 434–440.

Crawford, D. L., E. McCoy, J. M. Harkin, and P. Jones. 1973. Production of microbial protein from waste cellulose by *Thermomonospora fusca,* a thermophilic actinomycete. *Biotechnol. Bioeng.* **15:** 833–843.

Crawford, D. L., and J. B. Sutherland. 1979. The role of actinomycetes in the decomposition of lignocellulose. *Dev. Ind. Microbiol.* **20:** 143–151.

Crawford, D. L., J. B. Sutherland, A. L. Pometto, III, and R. L. Crawford. 1979a. Compositional changes in Douglas-fir phloem during biodegradation by *Streptomyces flavovirens,* presented at a symposium on Biosynthesis and Biodegradation of Cell Wall Components, ACS/CSJ Chem. Congress, April 1–6 (1979), Honolulu, Hawaii.

Crawford, R. L. 1975a. Novel pathway for degradation of protocatechuic acid in *Bacillus* species. *J. Bacteriol.* **121:** 531–536.

Crawford, R. L. 1975b. Degradation of 3-hydroxybenzoate by bacteria of the genus *Bacillus. Appl. Microbiol.* **30:** 439–444.

Crawford, R. L. 1975a. Pathways of 4-hydroxybenzoate degradation among species of *Bacillus. J. Bacteriol.* **127:** 204–210.

Crawford, R. L. 1976b. Degradation of homogentisate by strains of *Bacillus* and *Moraxella. Can. J. Microbiol.* **22:** 276–280.

Crawford, R. L. 1978. Hyroxylation of 4-hydroxyphenoxyacetate by a *Bacillus. FEMS Microbiol. Lett.* **4:** 233–234.

Crawford, R. L., J. Bromley, and P. E. Olson. 1979a. Catabolism of protocatechuate by *Bacillus macerans. Appl. Environ. Microbiol.* **37:** 614–618.

Crawford, R. L., and D. L. Crawford. 1978. Radioisotopic methods for the study of lignin biodegradation. *Dev. Ind. Microbiol.* **19:** 35–49.

Crawford, R. L., D. L. Crawford, C. Olofsson, L. Wikstrom, and J. M. Wood. 1977. Biodegradation of natural and man-made recalcitrant compounds with particular reference to lignin. *J. Agric. Food Chem.* **25:** 704–708.

Crawford, R. L., and T. D. Frick. 1977. Rapid spectrophotometric differentiation between glutathione-dependent and glutathione-independent gentisate and homogentisate pathways. *Appl. Environ. Microbiol.* **34:** 170–174.

Crawford, R. L., T. K. Kirk, and E. McCoy. 1975. Dissimilation of the lignin model compound veratrylglycerol-β-(o-methoxyphenyl) ether by *Pseudomonas acidovorans:* initial transformations. *Can. J. Microbiol.* **21:** 577–579.

Crawford, R. L., E. McCoy, J. M. Harkin, T. K. Kirk, and J. R. Obst. 1973. Degradation of methoxylated benzoic acids by a *Nocardia* from a lignin-rich environment: significance to lignin degradation and effect of chloro substituents. *Appl. Microbiol.* **26:** 176–184.

Crawford, R. L., E. McCoy, T. K. Kirk, and J. M. Harkin. 1973. Bacterial cleavage of an argylglycerol-β-aryl ether bond. *Appl. Microbiol.* **25:** 322–324.

Crawford, R. L., and P. E. Perkins (Olson). 1978a. Catabolism of 3,5-dihydroxybenzoate by *Bacillus brevis. FEMS Microbiol. Lett.* **4:** 161–162.

Crawford, R. L., and P. E. Perkins (Olson). 1978b. Microbial catabolism of vanillate: decarboxylation to guaiacol. *Appl. Environ. Microbiol.* **36:** 539–543.

Crawford, R. L., L. E. Robinson, and A. Cheh. 1979b. ^{14}C-Labeled lignins as substrates for the study of lignin biodegradation and transformation. In *Lignin Biodegradation: Microbiology, Chemistry, and Applications.* T. K. Kirk and T. Higuchi, eds. CRC Press, West Palm Beach, Fla., 1980.

DaCosta, E. W. B., and L. D. Bezemer. 1979. Some techniques for laboratory production of soft-rot in wood blocks for experimental purposes. *Holzforschung.* **33:** 7–10.

Dagley, S. 1971. Catabolism of aromatic compounds by microorganisms. *Adv. Microb. Physiol.* **6:** 1–46.

Dagley, S. 1977. Microbial degradation of organic compounds in the biosphere. *Surv. Prog. Chem.* **8:** 121–170.

Dagley, S., P. J. Chapman, and D. T. Gibson. 1963. Oxidation of β-phenylpropionic acid by an *Achromobacter. Biochim. Biophys. Acta.* **78:** 781–782.

Daugulis, A. J., and D. H. Bone. 1977. Submerged cultivation of edible white-rot fungi on tree bark. *Eur. J. Appl. Microbiol.* **4:** 159–166.

Davidson, R. W., W. A. Campbell, and D. J. Blaisdell. 1938. Differentiation of wood-decaying fungi by their reactions on gallic or tannic acid medium. *J. Agric. Res.* **57:** 683–695.

Davies, J. I., and W. C. Evans. 1964. Oxidative metabolism of naphthalene by soil pseudomonads: the ring-fission mechanism. *Biochem. J.* **91:** 251–261.

Dawes, E. A., and P. J. Senior. 1973. The role and regulation of energy reserve polymers in micro-organisms. *Adv. Microbiol. Physiol.* **10:** 135–267.

Day, W. C., M. J. Pelczar, Jr., and S. Gottlieb. 1949. The biological degradation of lignin. I. Utilization of lignin by fungi. *Arch. Biochem.* **23:** 360–369.

DeFrank, J. J., and D. W. Ribbons. 1977. *p*-Cymene pathway in *Pseudomonas putida:* ring cleavage of 2,3-dihydroxy-*p*-cumate and subsequent reactions. *J. Bacteriol.* **129:** 1365–1374.

Dekker, F. H., and G. N. Richards. 1973. Effect of delignification on the *in vitro* rumen digestion of polysaccharides of bagasse. *J. Sci. Fed. Agri.* **24:** 375–379.

Deschamps, A. M., G. Mahoudeau, and J. M. Lebeault. 1980. Fast degradation of kraft lignin by bacteria. *Eur. J. Appl. Microbiol. Biotechnol.* **9:** 45–51.

Dhawan, S., and J. K. Gupta. 1977. Enzymic hydrolysis of common cellulosic wastes by cellulase. *J. Gen. Appl. Microiol.* **23:** 155–161.

Drew, S. W., and K. L. Kadam. 1979. Lignin metabolism by *Aspergillus fumigatus* and white-rot fungi. *Dev. Ind. Microbiol.* **20:** 153–161.

Duncan, C. G. 1960. Wood-attacking capacities and physiology of soft-rot fungi. U.S.D.A. Forest Service Report 2173, 70 pages.

Eberhardt, G., and W. J. Schubert. 1956. Investigations on lignin and lignification. XVII. Evidence for the mediation of shikimic acid in the biogenesis of lignin building stones. *J. Am. Chem. Soc.* **78:** 2835–2837.

Effland, M. J. 1977. Modified procedure to determine acid-insoluble lignin in wood and pulp. *Tappi* **60:** 143–144.

Ellis, G. H., G. Matrone, and L. A. Maynard. 1946. A 72-percent H_2SO_4 method for the determination of lignin and its use in animal nutrition studies. *J. Anim. Sci.* **5:** 285–297.

Elmorsi, E. A., and D. J. Hopper. 1979. The catabolism of 5-hydroxyisophthalic acid by a soil bacterium. *J. Gen. Microbiol.* **111:** 145–152.

Endo, K., H. Kondo, and M. Ishimoto. 1977. Degradation of benzenesulfonate to sulfite in bacterial extract. *J. Biochem.* **82:** 1397–1402.

Enkvist, T., E. Solin, and U. Maunula. 1954. Studies on pine wood decayed by brown-rot. *Pap. Puu* **36:** 65–69.

Enoki, A., G. P. Goldsby, and M. H. Gold. 1980. Metabolism of the lignin model compounds veratrylglycerol-β-guaiacyl ether and 4-ethoxy-3-methoxyphenylglycerol-β-guaiacyl ether by *Phanerochaete chrysosporium*. *Arch. Microbiol.* **125:** 227–232.

Erdtman, H. 1933. Dehydrierungen in der Coniferylreihe. I. Dehydrodi-eugenol und Dehydrodiisoeugenol. *Biochem. Z.* **258:** 172–180.

Erickson, M., S. Larsson, and G. E. Miksche. 1973. Gas Chromatographische Analyse von Ligninoxydationsprodukten VII. *Acta Chem. Scand.* **27:** 127–140.

Erickson, M., and G. E. Miksche. 1974a. Two dibenzofurans obtained on oxidative degradation of the moss *Polytrichum communa* Hedw. *Acta Chem. Scand. B* **28:** 109–113.

Erickson, M., and G. E. Miksche. 1974b. Characterization of gymnosperm lignins by oxidative degradation. *Holzforschung* **28:** 135–138.

Erickson, M., and G. E. Miksche. 1974c. On the occurrence of lignin or polyphenols in some mosses and liverworts. *Phytochemistry* **13:** 2295–2299.

Eriksson, K.-E., P. Ander, B. Henningsson, T. Nilsson, and B. Goodell. 1976. Method for producing cellulose pulp. U.S. Patent 3,962,033.

Eriksson, K.-E., A. Grünewald, and L. Vallander. 1980. Studies of growth conditions in wood for three white-rot fungi and their cellulaseless mutants. *Biotechnol. Bioeng.* **22:** 363–376.

Eriksson, K.-E., and U. Lindholm. 1971. Ligninets mikrobiella nedbrytning. *Sven. Papperstid.* **74:** 701–706.

Esenther, G. R., and T. K. Kirk. 1974. Catabolism of aspen sapwood in *Reticulitermes flavipes* (Isoptera: Rhonitermidae). *Ann. Entomol. Soc. Am.* **67:** 989–999.

Eslyn, W. E., T. K. Kirk, and M. J. Effland. 1975. Changes in the chemical composition of wood caused by six soft-rot fungi. *Phytopathology* **65:** 473–475.

Evans, W. C. 1977. Biochemistry of the bacterial catabolism of aromatic compounds in anaerobic environments. *Nature* **270:** 17–22.

Evans, W. C., H. N. Fernley, and E. Griffiths. 1965. Oxidative metabolism of phenanthrene and anthracene by soil pseudomonads: the ring-fission mechanism. *Biochem. J.* **95:** 819–831.

Farmer, V. C., M. E. K. Henderson, and J. D. Russell. 1959. Reduction of certain aromatic acids to aldehydes and alcohols by *Polystictus versicolor*. *Biochim. Biophys. Acta* **35:** 202–211.

Farnham, R. S. 1978. Energy from peat: subcommittee 8 report to the Minnesota Energy Agency. Alternative Energy Research and Development Policy Formulation Project. St. Paul, Minn.

Faulkner, J. K., and D. Woodcock. 1968. The metabolism of phenylacetic acid by *Aspergillus niger*. *Phytochemistry.* **7:** 1741–1742.

Fenn, P., and T. K. Kirk. 1979. Ligninolytic system of *Phanerochaete chrysosporium*: inhibition by *o*-phthalate. *Arch. Microbiol.* **123:** 307–310.

Fergus, B. J., A. R. Procter, J. A. N. Scott, and D. A. I. Goring. 1969. The distribution of lignin in sprucewood as determined by ultraviolet microscopy. *Wood Sci. Technol.* **3:** 117–138.

Ferm, R., and A-C. Nilsson. 1969. Microbiological degradation of a commercial lignosulfonate. *Sven. Papperstidn.* **72:** 531–536.

Ferry, J. G., and R. S. Wolfe. 1976. Anaerobic degradation of benzoate to methane by a microbial consortium. *Arch. Microbiol.* **107:** 33–40.

Flaig, W., and K. Haider. 1961. Die Verwertung phenolischer Verbindungen durch Weibfalepilze. *Arch. Mikrobiol.* **40:** 212–223.

Floss, H. G., H. Guenther, D. Groeger, and D. Erge. 1969. Origin of the oxygen atoms in the conversion of anthranilic acid to 2,3-dihydroxybenzoic acid by *Claviceps paspali*. *Arch. Biochem. Biophys.* **131:** 319–324.

Ford, C. W. 1978. Effect of partial delignification on the *in vitro* digestibility of cell wall polysaccharides in *Digitaria decumbens* (Pangola grass). *Aust. J. Agric. Res.* **29:** 1157–1166.

Forrester, P. I., and G. M. Gaucher. 1972. Conversion of 6-methylsalicylic acid into patulin by *Penicillium urticae*. *Biochemistry* **11:** 1102–1107.

French, J. R. J., and D. E. Bland. 1975. Lignin degradation in the termites *Coptotermes lacteus* and *Nasutitermes exitiosus*. *Mater. Org.* **10:** 281–288.

Freudenberg, K. 1956. Lignin im Rahmen der polymeren Naturstoffs. *Angew. Chem.* **68:** 84–92.

Freudenberg, K. 1968. The constitution and biosynthesis of lignin. In *Constitution and Biosynthesis of Lignin*, Neish, A. C. and K. Freudenberg, eds. Springer-Verlag, New York, pp. 47–122.

Freudenberg, K., C. L. Chen, and G. Cardinale. 1962. Die Oxydation des methylierten natur-lichen und kunstlichen Lignins. *Chem. Ber.* **95:** 2814–2828.

Freudenberg, K., and L. Knof. 1957. Die Lignane des Fichtenholzes. *Chem. Ber.* **90:** 2857–2869.

Freudenberg, K., and A. C. Neish. 1968. *Constitution and Biosynthesis of Lignin*. Springer-Verlag, Berlin, pp. 82–97.

Fujikawa, N., and M. Ito. 1971. Meeting of the Japanese Society of Agricultural and Biological Chemists, p. 132 (cited in Harda and Watanabe, 1972).

Fukuzumi, T., and Y. Katayama. 1977. Bacterial degradation of dimer relating to structure of lignin. I. β-Hydroxypropiovanillone and coniferyl alcohol as initial degradation products from guaiacylglycerol-β-coniferylether by *Pseudomonas putida*. *Mokuzai Gakkaishi.* **23:** 214–215.

Fukuzumi, T., H. Takatuka, and K. Minami. 1969. Enzymic degradation of lignin. V. The effect of NADH on the enzymic cleavage of arylalkyl-ether bond in veratrylglycerol-β-guaiacylether as lignin model compound. *Arch. Biochem. Biophys.* **129:** 396–409.

Gadd, O. 1957. Wood decay resulting from rot fungi. *Paper and Timber* **8:** 363–374.

Ghosh, S., and D. L. Klass. 1979. Biomethanation of Minnesota reed sedge peat. *Rec. Recov. Conservat.* **4:** 115–139.

Gibson, D. T., J. R. Koch, and R. E. Kallio. 1968. Oxidative degradation of aromatic hydrocarbons by microorganisms. I. Enzymic formation of catechol from benzene. *Biochem. N.Y.* **7:** 2653–2662.

Gierer, J. 1970. The reactions of lignin during pulping—a description and comparison of conventional pulping processes. *Sven. Papperstidn.* **73:** 571–596.

Gierer, J., and A. E. Opara. 1973. Studies on the enzymatic degradation of lignin. The action of peroxidase and laccase on monomeric and dimeric model compounds. *Acta Chem. Scand.* **27:** 2909–2922.

Gold, M. H., and T. M. Cheng. 1978. Induction of colonial growth and replica plating of the white-rot basidiomycete *Phanerochaete chrysoporium*. *Appl. Environ. Microbiol.* **35:** 1223–1225.

Goldstein, I. S. 1975. Potential for converting wood into plastics. *Science* **189:** 847–852.

Gottlieb, S., and M. J. Pelczar, Jr. 1951. Microbiological aspects of lignin degradation. *Bacteriol. Rev.* **15:** 55–76.

Grabbe, K., R. Koenig, and K. Haider. 1968. Die Bildung der Phenol-oxydase und die Stoffwechselbeeinflussung durch Phenole bei *Polystictus versicolor*. *Arch. Mikrobiol.* **73:** 133–153.

Gradziel, K., K. Haider, J. Kochmanska, E. Malarczyk, and J. Trojanowski. 1978. Bacterial decomposition of synthetic ^{14}C-labeled lignin and lignin monomer derivatives. *Acta Microbiol. Pol.* **27:** 103–109.

Grant, D. J. W. 1971. Degradation of acetylsalicylic acid by a strain of *Acinetobacter lwoffi*. *J. Appl. Bacteriol.* **34:** 689–698.

Grisebach, H. 1977. Biochemistry of lignification. *Naturwissenschaften* **64:** 619–625.

Groeger, D., D. Erge, and H. G. Floss. 1965. Zur Biosynthese von 2,3-Dihydroxybenzosaure in Submerskutturen von *Claviceps paspali* Stevens et Hall. *Z. Naturforsch.* **20:** 856–858.

Grohn, H., and W. Deters. 1959. Uber den Abbau von Fichtenholz durch *Lenzites saepiaria*. *Holzforschung* **13:** 8–12.

Gross, G. G. 1977. Biosynthesis of lignin and related monomers. *Recent Adv. Phytochem.* **11:** 141–184.

Gross, G. G. 1979. Recent advances in the chemistry and biochemistry of lignin. *Recent Adv. Phytochem.* **12:** 177–220.

Gulyas, F. 1967. On the role played by several soil fungi in microbiological decomposition of lignin. *Agrokem. Talajtan* **16:** 137–150.

Gunnison, D., and M. Alexander. 1975. Basis for the resistance of several algae to microbial decomposition. *Appl. Microbiol.* **29:** 729–738.

Haars, A., and A. Huttermann. 1980. Macromolecular mechanism of lignin degradation by *Fomes annosus*. *Naturwissenschaften* **67:** 39–40.

Hackett, W. F., W. J. Connors, T. K. Kirk, and J. G. Zeikus. 1977. Microbial decomposition of synthetic ^{14}C-labeled lignins in nature: lignin biodegradation in a variety of natural materials. *Appl. Environ. Microbiol.* **33:** 43–51.

Hahlbrock, K., and H. Grisebach. 1979. Enzymic controls in the biosynthesis of lignin and flavonoids. *Ann. Rev. Plant Physiol.* **30:** 105–130.

Haider, K., and K. Grabbe. 1967. Die Rolle der Phenoloxydase bein Ligninabbau durch Weiss Foule-pilze. *Zentralbl. Bakteriol. Parasitenkd., Infectionskr., Hyg.* **205:** 91–96.

Haider, K., S. Lim, and W. Flaig. 1964. Experimente und Theorien über den Ligninabbau bei der Weiss-faule des Holzes und bei der Verrottung pflanzlicher Substanz in Boden. *Holzforschung* **18:** 81–88.

Haider, K., and J. P. Martin. 1975. Decomposition of specifically carbon-14 labeled benzoic and cinnamic acid derivaties in soil. *Soil Sci. Am. Proc.* **39:** 657–662.

Haider, K., and J. P. Martin. 1979. A comparison of the degradation of ^{14}C-labeled DHP and corn stalk lignins by micro- and macrofungi and by bacteria. In *Lignin Biodegradation: Microbiology, Chemistry, and Applications.* T. K. Kirk and T. Higuchi, eds. CRC Press, West Palm Beach, Fla., 1980.

Haider, K., J. P. Martin, and E. Rietz. 1977. Decomposition in soil of ^{14}C-labeled coumaryl alcohols: free and linked into dehydropolymer and plant lignins and model humic acids. *Soil Sci. Soc. Am.* **41:** 556–561.

Haider, K., and J. Trojanowski. 1975. Decomposition of specifically ^{14}C-labelled phenols and dehydropolymers. *Arch. Microbiol.* **105:** 33–42.

Haider, K., J. Trojanowski, and V. Sundman. 1978. Screening for lignin degrading bacteria by means of ¹⁴C-labelled lignins. *Arch. Microbiol.* **119:** 103–106.

Hajny, G. J., C. H. Gardner, and G. J. Ritter. 1951. Thermophilic fermentation of cellulose and lignocellulose materials. *Ind. Eng. Chem.* **43:** 1384–1389.

Hall, P. L., W. G. Glasser, and J. W. Drew. 1979. Enzymatic transformations of lignin. In *Lignin Biodegradation: Microbiology, Chemistry, and Applications.* T. K. Kirk and T. Higuchi, eds. CRC Press, West Palm Beach, Fla., 1980.

Hanson, K. R., and E. A. Havir. 1979. An introduction to the enzymology of phenylpropanoid biosynthesis. *Recent Adv. Phytochem.* **12:** 91–138.

Harada, T., and H. Watanabe. 1972. Formation of gentisic acid from salicylic acid and of protocatechuic acid from *m*- and *p*-hydroxybenzoic acids in medium containing glucose by a strain of *Trichoderma* sp. *J. Ferm. Technol.* **50:** 167–170.

Harelund, W. A., R. L. Crawford, P. J. Chapman and S. Dagley. 1975. Metabolic function and properties of 4-hyroxyphenylacetic acid 1-hydroxylase from *Pseudomonas acidovorans. J. Bacteriol.* **121:** 272–285.

Harkin, J. M. 1967. Lignin—a natural polymeric product of phenol oxidation. In *Oxidative Coupling of Phenols.* A. R. Battersby and W. I. Taylor, eds. Marcel Dekker, New York, pp. 243–321.

Harkin, J. M., D. L. Crawford, and E. McCoy. 1974. Bacterial protein from pulps and paper mill sludge. *Tappi* **57:** 131–134.

Harkin, J. M., and J. R. Obst. 1973a. Syringaldazine, an effective reagent for detecting laccase and peroxidase in fungi. *Experientia* **29:** 381–387.

Harkin, J. M., and J. R. Obst. 1937b. Lignification in trees: indication of exclusive peroxidase participation. *Science* **180:** 296–297.

Hasegawa, M., T. Higuchi, and H. Ishikawa. 1960. Formation of lignin in tissue culture of *Pinus strobus. Plant Cell Physiol.* **1:** 173–182.

Hashimoto, A. G., Y. R. Chen, and R. L. Prior. 1979. Methane and protein production from animal feedlot wastes. *J. Soil Water Conserv.* **Jan./Feb.:** 16–19.

Hashimoto, K. 1970. Oxidation of phenols by yeast. I. A new oxidation product from *p*-cresol by an isolated strain of yeast. *J. Gen. Appl. Microbiol.* **16:** 1–13.

Hashimoto, K. 1973. Oxidation of phenols by yeast. II. Oxidation of cresols by *Candida tropicalis. J. Gen. Appl. Microbiol.* **19:** 171–187.

Healy, J. B., and L. Y. Young. 1979. Anaerobic biodegradation of eleven aromatic compounds to methane. *Appl. Environ. Microbiol.* **38:** 84–89.

Hemmingson, J. A. 1979. A new way of forming lignin-carbohydrate bonds. Etherification of model benzyl alcohols in alcohol/water mixtures. *Aust. J. Chem.* **32:** 225–229.

Henderson, M. E. K. 1957. Metabolism of methoxylated aromatic compounds by soil fungi. *J. Gen. Microbiol.* **16:** 686–695.

Henderson, M. E. K. 1961. The metabolism of aromatic compounds related to lignin by some hyphomycetes and yeast-like fungi of soil. *J. Gen. Microbiol.* **26:** 155–165.

Henderson, M. E. K. 1968. Enrichment in soil fungi which utilize aromatic compounds. *Plant Soil* **23:** 339–350.

Hergert, H. L. 1971. Infrared spectra. In *Lignins: Occurrence, Formation, Structure, and Reactions.* K. V. Sarkanen and C. H. Ludwig, eds. Wiley-Interscience, New York. p. 280.

Higashi, T., and Y. Sakamoto. 1960. Oxidation of anthranilic acid catabolized by *Pseudomonas* cell-free extract. *J. Biochem. (Tokyo)* **48:** 147–149.

Higuchi, T. 1979. Microbial degradation of dilignols as lignin models. In *Lignin Biodegradation:*

Microbiology, Chemistry, and Applications. T. K. Kirk and T. Higuchi, eds. CRC Press, West Palm Beach, Fla., 1980.

Higuchi, T., I. Kawamura, and H. Kawamura. 1955. Properties of the lignin in decayed wood. *J. Jpn. Forest Soc.* **37:** 298–302.

Higuchi, T., M. Shimada, F. Nakatsubo, and M. Tanahashi. 1977. Differences in biosynthesis of guaiacyl and syringyl lignins in woods. *Wood Sci. Technol.* **11:** 153–167.

Hiroi, T., and K.-E. Eriksson. 1976. Influence of cellulose on the degradation of lignins by the white-rot fungus *Pleurotus ostreatus. Sven. Papperstdn.* **79:** 157–161.

Hobbie, J. E., and C. C. Crawford. 1969. Respiration corrections for bacterial uptake of dissolved organic compounds in natural waters. *Limnol. Oceanogr.* **14:** 528–532.

Hopkinson, C. S. Jr., and J. W. Day, Jr. 1980. Net energy analysis of alcohol production from sugarcane. *Science* **207:** 302–304.

Humphrey, A. E., A. Moreira, W. Armiger, and D. Zabriskie. 1977. Production of single cell protein from cellulose wastes. *Biotechnol. Bioeng. Symp. No. 7,* 45–64.

Ishikawa, H., W. J. Schubert, and F. F. Nord. 1963a. Investigations on lignins and lignification. XXVII. The enzymic degradation of softwood lignin by white-rot fungi. *Arch. Biochem. Biophys.* **100:** 131–139.

Ishikawa, H., W. J. Schubert, and F. F. Nord. 1963b. Investigations on lignins and lignification. XXVIII. The degradation by *Polyporus versicolor* and *Fomes fomentarius* of aromatic compounds structurally related to softwood lignin. *Arch. Biochem. Biophys.* **100:** 140–149.

Isono, M. 1958. Degradation of phenylacetic acid by *Penicillium chrysogenum. J. Agric. Chem. Soc.* **32:** 256–259.

Iwahara, S. 1979. Microbial degradation of DHP. In *Lignin Biodegradation: Microbiology, Chemistry, and Applications.* T. K. Kirk and T. Higuchi, eds. CRC Press, Cleveland, Ohio.

Iwahara, S., and T. Higuchi. 1979. Microbial degradation of lignin model compounds. Presented at a Symposium on Biosynthesis and Biodegradation of Cell Wall Components. ACS/CSJ Chemical Congress, April 1–6, Honolulu, Hawaii.

Iwahara, S., M. Kuwahara, and I. Higuchi. 1977. Microbial degradation of dehydrogenation polymer of coniferyl alcohol (DHP). *Hakkokogaku Kaishi.* **55:** 325–329. (in Japanese)

Jaschof, H. 1964. Preliminary studies of the decomposition of lignin by bacteria isolated from lignite. *Geochim. Cosmochim. Acta* **28:** 1623.

Jefferies, T. W., D. E. Eveleigh, J. D. MacMillan, F. W. Parrish, and E. T. Reese. 1977. Enzymatic hydrolysis of the walls of yeast cells and germinated fungal spores. *Biochim. Biophys. Acta* **499:** 10–23.

Jensen, V. 1974. Decomposition of angiosperm tree leaf litter, In *Biology of Plant Litter Decomposition,* Vol. 1. C. H. Dickinson and G. J. F. Pugh, eds. Academic Press, New York. 1974.

Johnson, B. F., and R. Y. Stanier. 1971. Dissimilation of aromatic compounds by *Alcaligenes eutrophus. J. Biol. Chem.* **107:** 468–475.

Johnson, D. B., W. E. Moore, and L. C. Zank. 1961. The spectrophotometric determination of lignin in small wood samples. *Tappi* **44:** 793.

Kalahatai, K. K., A. M. D. Nambudiri, and A. A. Kaarik. 1974. Decomposition of wood. In *Biology of Plant Litter Decomposition,* Vol. 1. C. H. Dickinson and G. J. F. Pugh, eds. Academic Press, New York, 1974.

Kalghatgi, K. K., A. M. D. Nambudiri, J. V. Bhat, and P. V. Subba Rao. 1974. Degradation of L-phenylalanine by *Rhizoctania solani. Ind. J. Biochem. Biophys.* **11:** 116–118.

Kaplan, D. 1979. Reactivity of different oxidases with lignins and lignin model compounds. *Phytochemistry* **18:** 1917–1920.

Kaplan, D., and R. Hartenstein. 1978. Studies on monooxygenases and dioxygenases in soil invertebrates and bacterial isolates from the gut of the terrestrial isopod *Oniscus asellus* L. *Comp. Biochem. Physiol* **60 B:** 47–50.

Kaplan, D. L., and R. Hartenstein. 1980. Decomposition of lignins by microorganisms. *Soil Biol. Biochem.* **12:** 65–74.

Katagiri, M., H. Maeno, S. Yamamoto, and O. Hayaishi. 1965. Salicylate hydroxylase, a monooxygenase requiring flavin adenine dinucleotide. *J. Biol. Chem.* **240:** 3414–3417.

Katayama, Y., and T. Fukuzumi. 1978. Bacterial degradation of dimers structurally related to lignin. II. Initial intermediate products from dehydrodiconiferyl alcohol by *Pseudomonas putida. Mokuzai Gakkaishi* **24:** 643–649.

Katayama, Y., and T. Fukuzumi. 1979. Bacterial degradation of dimers structurally related to lignin. III. Metabolism of α-veratryl-β-guaiacylpropionic acid and D,L-pinoresinol by *Pseudomonas putida. Mokuzai Gakkaishi* **25:** 67–76.

Kawakami, H. 1975a. Bacterial degradation of lignin model compounds. III. On the degradation of model dimers and 5-position condensed type compounds of guaiacyl nuclei. *Mokuzai Gakkaishi* **21:** 629–634.

Kawakami, H. 1975b. Bacterial degradation of lignin model compounds. I. On the cleavage of aromatic nuclei. *Mokuzai Gakkaishi* **21:** 93–100.

Kawakami, H. 1976a. Bacterial degradation of lignin model compounds. IV. On the aromatic ring cleavage of vanillic acid. *Mokuzai Gakkaishi* **22:** 246–251.

Kawakami, H. 1976b. Bacterial degradation of lignin. I. Degradation of MWL by *Pseudomonas ovalis. Mokuzai Gakkaishi* **22:** 252–257.

Kawakami, H., N. Mori, and T. Kanda. 1975a. Biodegradation of compounds of pulp waste effluents by bacteria. 2. On the degradation of lignin sulfonates. *Jpn. Tappi* **29:** 596–601.

Kawakami, H., M. Sugiura, and T. Kanda. 1975b. Biodegradation of components of pulp waste effluents by bacteria. 1. On the degradation of kraft lignin. *Jpn. Tappi* **29:** 309–315.

Kawase, K. 1962. Chemical compounds of wood decayed under natural conditions and their properties. *J. Fac. Agric. Hokkaido Imp. Univ.* **52:** 186–245.

Keith, C. L., R. L. Bridges, L. R. Fina, K. L. Iverson, and J. A. Cloran. 1978. The anaerobic decomposition of benzoic acid during methane fermentation. IV. Dearomatization of the ring and volatile fatty acids formed on ring rupture. *Arch. Microbiol.* **118:** 173–176.

Kennedy, S. I. T., and C. A. Fewson. 1966. The oxidation of mandelic acid and related compounds by organism NCIB 8250. *Biochem. J.* **100:** 25p.–26p.

Keyser, P., T. K. Kirk, and J. G. Zeikus. 1978. Lignolytic enzyme system of *Phanerochaete chrysosporium:* synthesized in the absence of lignin in response to nitrogen starvation. *J. Bacteriol* **135:** 790–797.

King, B., and H. O. W. Eggins. 1977. Micromorphology of streptomycete colonization of wood. *J. Inst. Wood Sci.* **7:** 24–49.

Kirk, T. K. 1971. Effects of microorganisms on lignin. *Ann. Rev. Phytopathol.* **9:** 185–210.

Kirk, T. K. 1973a. Polysaccharide integrity as related to the degradation of lignin in wood by white-rot fungi. *Phytopathology* **63:** 1504–1507.

Kirk, T. K. 1973b. *Wood Deterioration and Its Prevention by Preservative Treatments.* Syracuse University Press, Syracuse, N. Y., pp. 149–181.

Kirk, T. K. 1975. Effects of a brown-rot fungus, *Lenzites trabea,* on lignin in spruce wood. *Holzforschung* **29:** 99–107.

Kirk, T. K., and E. Adler. 1970. Methoxyl-deficient structural elements in lignin of sweetgum decayed by a brown-rot fungus. *Acta Chem. Scand.* **24:** 3379–3390.

Kirk, T. K., and H-M. Chang. 1974. Decomposition of lignin by white-rot fungi. I. Isolation of heavily degraded lignins from decayed spruce. *Holzforschung* **28:** 217–222.

Kirk, T. K., and H-M. Chang. 1975. Decomposition of lignin by white-rot fungi. II. Characterization of heavily degraded lignins from decayed spruce. *Holzforschung* **29:** 56–64.

Kirk, T. K., W. J. Connors, R. D. Bleam, W. F. Hackett, and J. G. Zeikus. 1975. Preparation and microbial decomposition of synthetic [^{14}C] lignins. *Proc. Natl. Acad. Sci. USA* **72:** 2515–2519.

Kirk, T. K., W. J. Connors, and J. G. Zeikus. 1976. Requirement for a growth substrate during lignin decomposition by two wood-rotting fungi. *Appl. Environ. Microbiol.* **32:** 192–194.

Kirk, T. K., W. J. Connors, and J. G. Zeikus. 1977. Advances in understanding the microbiological degradation of lignin. *Rev. Adv. Phytopathol.* **11:** 369–394.

Kirk, T. K., and P. Fenn. 1979. Formation and action of the ligninolytic system in Basidiomycetes. *Proc. Symp. Brit. Mycol. Soc.* In press.

Kirk, T. K., and T. L. Highley. 1973. Quantitative changes in structural components of conifer woods during decay by white-and brown-rot fungi. *Phytopathology* **63:** 1338–1342.

Kirk, T. K., T. Higuchi, and H-M. Chang, eds. *Lignin Biodegradation: Microbiology, Chemistry, and Applications.* CRC Press, West Palm Beach, Fla., 1980.

Kirk, T. K., S. Larsson, and G. E. Miksche. 1970. Aromatic hydroxylation resulting from attack of lignin by a brown-rot fungus. *Acta Chem. Scand.* **24:** 1470–1472.

Kirk, T. K., and K. Lundquist. 1970. Comparison of sound and white-rotted sapwood of sweetgum with respect to properties of the lignin and composition of extractives. *Sven. Papperstidn.* **73:** 294–306.

Kirk, T. K., and W. E. Moore. 1972. Removing lignin from wood with white-rot fungi and digestibility of resulting wood. *Wood and Fiber* **4:** 72–79.

Kirk, T. K., E. Schultz, W. J. Connors, L. F. Lorenz, and J. G. Zeikus. 1978. Influence of culture parameters on lignin metabolism by *Phanerochaete chrysosporium. Arch. Microbiol.* **117:** 277–285.

Kirk, T. K., and H. H. Yang. 1979. Partial delignification of unbleached kraft pulp with ligninolytic fungi. *Biotechnol. Lett.* **1:** 347–352.

Kishore, G., M. Sugumaran, and C. S. Vaidyanathan. 1976. Metabolism of DL-(±)-phenylalanine by *Aspergillus niger. J. Bacteriol.* **128:** 182–191.

Kiyohara, H., and K. Nagao. 1977. Enzymic conversion of 1-hydroxy-2-naphthoate in phenanthrene-grown *Aeromonas* sp. S45P1. *Agric. Biol. Chem.* **41:** 705–707.

Kiyohara, H., and K. Nagao. 1978. The catabolism of phenanthrene and naphthalene by bacteria. *J. Gen. Microbiol.* **105:** 69–75.

Klason, P. (cited in Adler, 1972). Bericht uber die Hauptversammlung des vereins der Zellstoff und Papierchemiker, p. 52.

Kohmoto, K., S. Nishimura, and I. Hiroe. 1970. Pathochemical studies on *Rhizoctonia* disease. I. *meta*-Hydroxylation of phenylacteic acid by *Rhizoctonia solani. Phytopathology* **60:** 1025–1026.

Konetzka, W. A., M. J. Pelczar, and S. Gottlieb. 1952. The biological degradation of lignin. III. Bacterial degradation of alpha-conidendrin. *J. Bacteriol.* **63:** 771–778.

Kosikova, B., D. Joniak, and L. Kosakova. 1979. On the properties of benzyl ether bonds in the lignin-saccharidic complex isolated from spruce. *Holzforschung* **33:** 11–14.

Kratzl, K., and G. Billek. 1957. Synthesis and testing of lignin precursors. *Tappi* **40:** 269–285.

Krisnangkura, K., and M. H. Gold. 1979. Characterization of guaiacyl lignin degraded by the white rot basidiomycete *Phanerochaete chrysosporium. Holzforschung* **33:** 174–176.

Krisnangkura, K., M. B. Mayfield, M. H. Gold, K. Moore, and T. K. Kirk. 1979. Regulation of lignin model compound metabolism by *Phanerochaete chrysosporium*. Presented at a Symposium on Biosynthesis and Biodegradation of Cell Wall Components. ACS/CSJ Chemical Congress, April 1–6, Honolulu, Hawaii.

Kunita, N. 1955. (cited in Sugumaran and Vaidyanathan, 1978). *Med. J. Osaka Univ.* **6:** 703.

Kuroda, H., and T. Higuchi. 1976. Characterization and biosynthesis of mistletoe lignin. *Phytochemistry* **15:** 1511–1514.

Kyrklund, B., and G. Strandell. 1967. A modified chlorine number for evaluation of the cooking degree of high-yield pulps. *Papper och Tra* **49:** 99–106.

Lack. L. 1959. The enzymic oxidation of gentisic acid. *Biochim. Biophys. Acta.* **34:** 117–123.

Lai, Y. Z., and K. V. Sarkanen. 1971. Isolation and structural studies. In *Lignin: Occurrence, Formation, Structure, and Reactions.* Wiley-Interscience, New York, pp. 165–240.

Larway, P., and W. C. Evans. 1965. Metabolism of quinol and resorcinol by soil pseudomonads. *Biochem J.* **95:** 52.

Law, L. 1959. A role of the laccase of wood-rotting fungi. *Physiol. Plant.* **12:** 854–861.

Ledingham, G. A., and G. A. Adams. 1942. Biological decomposition of chemical lignin. II. Studies on the decomposition of Ca-lignosulfonate by wood destroying and soil fungi. *Can. J. Res. Soc. C* **20:** 13–27.

Lee, K. E., and T. G. Wood. 1971. *Termites and Soils.* Academic Press, London.

Lee, Y. W., and J. M. Pepper. 1978. Lignin and related compounds. VII. The isolation of a trimeric lignin compound by the hydrogenolysis of spruce wood. *Tetrahedron Lett.* **51:** 5061–5062.

Leopold, B. 1951. Studies on lignin. VIII. Nitrobenzene oxidation and sulphonation of wood decayed by brown-rotting fungi. *Sven Kem. Tidskr.* **63:** 260–271.

Levi, M. P., and R. D. Preston. 1965. A chemical and microscopic examination of the action of the soft-rot fungus *Chaetomium globosum* on beechwood (*Fagus sylv.*) *Holzforschung* **19:** 183–190.

Light, R. J. 1969. 6-Methylsalicylic acid decarboxylase from *Penicillium patulum*. *Biochim. Biophys. Acta* **191:** 430–438.

Light, R. J., and G. Vogel. 1974. Patulin biosynthesis: the role of mixed-function oxidases in the hydroxylation of *m*-cresol. *Eur. J. Biochem.* **49:** 443–455.

Lofty, J. R. 1974. Oligochaetes. In *Biology of Plant Litter Decomposition,* Vol. 2. C. H. Dickinson and G. J. F. Pugh, eds. Academic Press, New York, pp. 467–488.

Ludemann, H.-D., and H. Nimz. 1973. Carbon-13 nuclear magnetic resonance spectra of lignins. *Biochem. Biophys. Res. Commun.* **52:** 1162–1169.

Lundquist, K. 1979a. NMR studies of lignins. 2. Interpretation of the ^{1}H NMR spectrum of acetylated birch lignin. *Acta Chem. Scand. B* **33:** 27–30.

Lundquist, K. 1979b. NMR studies of lignins. 3. ^{1}H NMR spectroscopic data for lignin model compounds. *Acta Chem. Scand. B* **33:** 418–420.

Lundquist, K., and T. K. Kirk. 1978. *De novo* synthesis and decomposition of veratryl alcohol by a lignin-degrading basidiomycete. *Phytochemistry* **17:** 1676.

Lundquist, K., and T. K. Kirk. 1980. Fractionation-purification of an industrial kraft lignin. *Tappi* **63:** 80–82.

Lundquist, K., T. K. Kirk, and W. J. Connors. 1977. Fungal degradation of kraft lignin and lignin sulfonates prepared from synthetic ^{14}C-lignins. *Arch. Microbiol.* **112:** 291–296.

Lundquist, K., and R. Simonson. 1975. Lignin preparations with very low carbohydrate content. *Sven. Papperstidn.* **78:** 390.

Mandels, M., J. Kostick, and R. Darizek. 1971. The use of adsorbed cellulose in the continuous conversion of cellulose to glucose. *J. Polym. Sci. Part C* **36:** 445–459.

Martin, J. P., and K. Haider. 1977. Decomposition in soil of specifically ^{14}C-labeled DHP and cornstalk lignins, model humic acid-type polymers and coniferyl alcohols. In Shelagh Freeman, ed. *Proc. Int. Symp. of Soil Organic Matter Studies,* September 6–10, 1976, Braunschweig, Germany. Int. At. Ener. Agency and Agrochim.

Martin, J. P., and K. Haider. 1979. Biodegradation of ^{14}C-labeled model and cornstalk lignins, phenols, model phenolase humic polymers, and fungal melanins as influenced by a readily available carbon source and soil. *Appl. Environ. Microbiol.* **38:** 283–289.

Mellenberger, R. W., L. D. Stratter, M. A. Millett, and M. A. Baker. 1971. Digestion of aspen, alkali-treated aspen, and aspen bark by goats. *J. Anim. Sci.* **32:** 756–763.

Migita, W., and I. Kawamura. 1944. Chemical analyses of wood. *J. Agric. Chem. Soc. Jpn.* **20:** 348.

Miksche, G. E., and S. Yasuda. 1978. Lignin of "giant" mosses and some related species. *Phytochemistry* **17:** 503–504.

Millett, M. A., A. J. Baker, W. E. Feist, R. W. Mellenberger, and L. D. Stratter. 1970. Modifying wood to increase its *in vitro* digestibility. *J. Anim. Sci.* **31:** 781–788.

Moore, W. E., and D. B. Johnson. 1967. Procedures for the analysis of wood and wood products. USDA Forest Service, Forest Products Laboratory, Madison, Wisc.

Moore, K., and G. H. N. Towers. 1967. Degradation of aromatic amino acids by fungi. 1. Fate of L-phenylalanine in *Schizophyllum commune. Can. J. Biochem.* **45:** 1659–1665.

Moo-Young, M., A. J. Daugulis, D. S. Chahal, and D. G. Macdonald. 1979. The Waterloo Process for SCP production from waste biomass. *Proc. Biochem.* **14:** 38–40.

Muranaka, M., S. Kinoshita, Y. Yamada, and H. Okada. 1976. Decomposition of lignin model compound, 3-(2-methoxy-4-formylphenoxy)-1,2-propanediol by bacteria. *J. Ferment. Technol.* **54:** 635–639.

Murphy, G., G. Vogel, E. Krippahl, and F. Lynen. 1974. Patulin biosynthesis: the role of mixed-function oxidases in the hydroxylation of *m*-cresol. *Eur. J. Biochem.* **49:** 443–455.

Murray, K., C. T. Duggleby, J. M. Sala-Trepat, and P. A. Williams. 1972. The metabolism of benzoate and methylbenzoates via the *meta*-cleavage pathway by *Pseudomonas arvilla* mt-2. *Eur. J. Biochem.* **28:** 310–310.

Nakamura, Y., and T. Higuchi. 1976. Ester linkage of *p*-coumaric acid in bamboo lignin. *Holzforschung* **30:** 187–191.

Neuhauser, E., and R. Hartenstein. 1976a. Degradation of phenol, cinnamic acid and quinic acid in the terrestrial crustacean *Oniscus asellus. Soil Biol Biochem.* **8:** 95–98.

Neuhauser, E., and R. Hartenstein. 1976b. On the presence of O-demethylase activity in invertebrates. *Comp. Biochem. Physiol* **53:** 37–39.

Neuhauser, E. F., R. Hartenstein, and W. J. Connors. 1978. Soil invertebrates and the degradation of vanillin, cinnamic acid, and lignins. *Soil. Biol. Biochem.* **10:** 431–435.

Neuhauser, E., C. Younell, and R. Hartenstein. 1974. Degradation of benzoic acid in the terrestrial crustacean *Oniscus asellus. Soil. Biol. Biochem.* **6:** 101–107.

Neujahr, H. Y., and A. Gaal. 1973. Phenol hydroxylase from yeast. Purification and properties of the enzyme from *Trichosporon cutaneum. Eur. J. Biochem.* **35:** 386–400.

Neujahr, H. Y., and A. Gaal. 1975. Phenol hydroxylase from yeast. Sulfhydryl groups in phenol hydroxylase from *Trichosporon cutaneum. Eur. J. Biochem.* **58:** 351–357.

Neujahr, H. Y., and J. M. Varga. 1970. Degradation of phenols by intact cells and cell-free preparations of *Trichosporon cutaneum. Eur. J. Biochem.* **13:** 37–44.

Nilsson, T., and J. Ginns. 1979. Cellulolytic activity and the taxonomic position of selected brown-rot fungi. *Mycologia* **71:** 170–177.

Nimz, H. 1969. Uber ein neues Abbauverfahren des Lignins. *Chem. Ber.* **102:** 799–810.

Nimz, H. 1974. Beech lignin—proposal of a constitutional scheme. *Angew. Chem.* **86:** 336–344.

Nimz, H., J. Ebel, and H. Grisebach. 1975. On the structure of lignin from soybean cell suspension cultures. *Z. Naturforsch.* **30:** 442–444.

Nimz, H., and H.-D. Ludemann. 1976. Carbon-13 NMR spectra of lignins. 6. Lignin and DHP acetates. *Holzforschung* **30:** 33–40.

Nimz, H., I. Mogharab, and H.-D. Ludemann. 1974. ^{13}C-kernresonanzspektren von Ligninen. 3. Vergleich von Fichtenlignin mit kunstlichem Lignin nach Freudenberg. *Makromol. Chem.* **175:** 2563–2575.

Nishikada, T. 1951. Isolation of o-hydroxyphenylacetic acid from the penicillin impurities. *Antibiotic (Tokyo)* **41:** 299–300.

Nord, F. F. 1965. Biochemistry of lignin. In *Holz und Organismen.* Int. Symp. Berlin-Dahlem. G. Becker and W. Liese, eds., Duncker und Humblot, Berlin, 1966, pp. 163–171.

Nord, F. F., and J. Schubert. 1955. Enzymatic studies on cellulose, lignin and the mechanism of lignification. *Holzforschung* **5:** 1–9.

Odier, E., and B. Monties. 1977. Activité ligninolytique in vitro de Bacteries isolees de paille de Ble en decomposition. *C.R. Acad. Sci. Paris, Serie D* **284:** 2175–2178.

Odier, E., and B. Monties. 1978a. Biodegradation of wheat lignin by *Xanthomonas* 23. *Ann. Microbiol. (Inst. Pasteur)* **129A:** 361–377.

Odier, E., and B. Monties. 1978b. Biodegradation des Lignines par les Bacteries. *C.-R. J. Int. D'Etude de L'assemb. Gen. Nancy.* Mai 17–19: 207–213.

Ohta, M., T. Higuchi, and S. Iwahara. 1979. Microbial degradation of dehydrodiconiferyl alcohol, a lignin substructure model. *Arch. Microbiol.* **121:** 23–28.

Ornston, L. N. 1966a. The conversion of catechol and protocatechaute to β-ketoadipate by *Pseudomonas putida.* II. Enzymes of the protcatechuate pathway. *J. Biol. Chem.* **241:** 3787–3794.

Ornston, L. N. 1966b. The conversion of catechol and protocatechuate to β-ketoadipate by *Pseudomonas putida.* III. Enzymes of the catechol pathway. *J. Biol. Chem.* **241:** 3795–3799.

Ornston, L. N. 1966c. The conversion of catechol and protocatechuate to β-ketoadipate by *Pseudomonas putida.* IV. Regulation. *J. Biol. Chem.* **241:** 3800–3810.

Ornston, L. N., and R. Y. Stanier. 1966. The conversion of catechol and protocatechuate to β-ketoadipate by *Pseudomonas putida.* I. Biochemistry. *J. Biol. Chem.* **241:** 3776–3786.

Pamment, N., C. W. Robinson, and M. Moo-Young. 1979. Pulp and paper mill solid wastes as substrates for single-cell protein production. *Biotechnol. Bioeng.* **21:** 561–573.

Pearl, I. A. 1967. *The Chemistry of Lignin.* Marcel Dekker, New York.

Peitersen, N. 1975. Cellulase and protein production from mixed culture of *Trichoderma viride* and a yeast. *Biotechnol. Bioeng.* **17:** 1291–1299.

Pelczar, M. J. Jr., S. Gottlieb, and W. C. Day. 1950. Growth of *Polyporus versicolor* in a medium with lignin as the sole carbon source. *Arch. Biochem.* **25:** 449–451.

Pew, J. C., and P. Weyna. 1962. Fine grinding, enzyme digestion, and the lignin-cellulose bond in wood. *Tappi* **45:** 247–256.

Phelan, M. B., D. L. Crawford, and A. L. Pometto, III. 1979. Isolation of lignocellulose-decomposing actinomycetes and degradation of specifically ^{14}C-labeled lignocellulose by six selected *Streptomyces* strains. *Can. J. Microbiol* **25:** 1270–1276.

Pickett-Heaps, J. D. 1968. Xylem wall deposition: radioautographic investigations using lignin precursors. *Protoplasma* **65:** 181–205.

Power, D. M., G. H. N. Towers, and A. C. Neish. 1965. Biosynthesis of phenolic acids by certain wood-destroying basidiomycetes. *Can. J. Biochem.* **43:** 1397–1407.

Que, L. 1978. Extradiol cleavage of *o*-aminophenol by pyrocatechase. *Biochem. Biophys. Res. Commun.* **84:** 123–129.

Quieryzy, P., N. Therien, and A. Leduy. 1979. Production of *Candida utilis* protein from peat extracts. *Biotechnol. Bioeng.* **21:** 1175–1190.

Räiha, M., and V. Sundman. 1975. Characterization of lignosulfonate-induced phenol oxidase activity in the atypical white-rot fungus *Polyporus dichrous*. *Arch. Microbiol.* **105:** 73–76.

Ramachandran, A., V. Subramanian, M. Sugumaran, an C. S. Vaidyanathan. 1979. Purification and properties of pyrocatechuate decarboxylase from *Aspergillus niger*. *FEMS Microbiol. Lett.* **5:** 421–425.

Ramanarayanan, M., and C. S. Vaidyanathan. 1975. Formation of 2,4-dihydroxybenzoic acid and resorcinol as intermediates in the degradation of salicylic acid by *Aspergillus nidulans*. *Ind. J. Exp. Biol.* **13:** 393–396.

Ramasamy, K., K. Prakasam, J. Bevers, and H. Verachtert. 1979. Production of bacterial proteins from cellulosic materials. *J. Appl. Bacteriol.* **46:** 117–124.

Reid, I. D. 1979. The influence of nutrient balance on lignin degradation by the white-rot fungus *Phanerochaete chrysosporium*. *Can. J. Bot.* **57:** 2050–2058.

Ribbons, D. W. 1966. Metabolism of *o*-cresol by *Pseudomonas aeruginosa* T$_1$. *J. Gen. Microbiol.* **44:** 221–231.

Ribbons, D. W. 1970. Stoichiometry of *o*-demethylase activity in *Pseudomonas aeruginosa*. *FEBS Lett.* **8:** 101–104.

Ribbons, D. W., and W. C. Evans. 1960. Oxidative metabolism of phthalic acid by soil pseudomonads. *Biochem. J.* **76:** 310–318.

Ribbons, D. W., and P. J. Senior. 1970. 2,3-Dihydroxybenzoate 3,4-oxygenase from *Pseudomonas flourescens*—oxidation of a substrate analog. *Arch. Biochem. Biophys.* **138:** 557–565.

Robbins, J. E., M. T. Armold, and S. L. Lacher. 1979. Methane production from cattle waste and delignified straw. *Appl. Environ. Microbiol.* **38:** 175–177.

Robinson, L. E., and R. L. Crawford, 1978. Degradation of ^{14}C-labeled lignins by *Bacillus megaterium*. *FEMS Microbiol. Lett.* **4:** 301–302.

Rösch, R. 1965. Über die Funktion der Phenolyxadasen holzabbauender Pilze. In *Holz und Organismen*, Int. Symp. Berlin-Dahlem, G. Becker and W. Liese, eds. Duncker und Humblot, Berlin; 1966, pp. 173–185.

Sakakibara, A. 1980. A structural model of softwood lignin. *Wood Sci. Technol.* **14:** 89–100.

Sala-Trepat, J. M., and W. C. Evans. 1971. The *meta* cleavage of catechol by *Azotobacter* species: 4-oxalocrotonate pathway. *Eur. J. Biochem.* **20:** 400–413.

Sala-Trepat, J. M., K. Murray, and P. A. Williams. 1972. The metabolic divergence in the *meta* cleavage of catechols by *Pseudomonas putida* NCIB 10015: physiological significance and evolutionary implications. *Eur. J. Biochem.* **28:** 347–356.

Saleh, T. M., L. Leney, and K. V. Sarkanen. 1967. Radioautographic studies of cottonwood, douglas fir and wheat plants. *Holzforschung* **21:** 116–119.

Sarkanen, K. V., H.-M. Chang, and G. G. Allan. 1967. Species variation in lignins. III. Hardwood lignins. *Tappi.* **50:** 587–590.

Sarkanen, K. V., and H. L. Hergert. 1971. Classification and distribution. In *Lignins: Occurrence, Formation, Structure, and Reactions*. K. V. Sarkanen and C. H. Ludwig, eds., Wiley-Interscience, New York, pp. 43–94.

Sarkanen. K. V., and C. H. Ludwig, eds. *Lignins: Occurrence, Formation, Structure, and Reactions.* Wiley-Interscience, New York, 1971.

Savory, J. G. 1954. Breakdown of timber by ascomycetes and Fungi Imperfecti. *Ann. Appl. Biol.* **41:** 336–347.

Savory. J. G., and L. C. Pinion. 1958. Chemical aspects of decay of beech wood by *Chaetomium globosum. Holzforschung* **12:** 99–103.

Schmid, L. A. 1975. Feedlot wastes to useful energy—fact or fiction. *J. Environ. Eng. Div., Proc. Am. Soc. Civil Eng.* **101:** 787–794.

Schmidt, O. 1978. On the bacterial decay of the lignified cell wall. *Holzforschung* **32:** 214–215.

Schubert, W. J., and F. F. Nord. 1950. Investigations on lignin and lignification. II. The characterization of enzymatically liberated lignin. *J. Am. Chem. Soc.* **72:** 3835–3838.

Scott, A. I., and L. C. Beadling. 1974. Biosynthesis of patulin. Dehydrogenase and dioxygenase enzymes of *Penicillium patulum. Bioorg. Chem.* **3:** 281–301.

Scott, W. W., E. B. Fred, and W. H. Peterson. 1930. Products of the thermophilic fermentation of cellulose. *Ind. Eng. Chem.* **22:** 731–735.

Scott, A. I., L. Zamir, G. T. Phillips, and M. Yalpani. 1973. The biosynhthesis of patulin. *Bioorg. Chem.* **2:** 124–139.

Seidman, M. M., A. Toms, and J. M. Wood. 1969. Influence of side-chain substituents on the position of cleavage of the benzene ring by *Pseudomonas flourescens. J. Bacteriol.* **97:** 1192–1197.

Seifert, K. 1966. Die chemische Veränderung der Buchenholz-Zellwand durch Moder fäule (*Chaetomium globosum* Kunze). *Holzforschung* **24:** 185–189.

Seifert, K., and G. Becker. 1965. Der chemische Abbau von Laubund Nadelholzarten durch verschieden Termiten. *Holzforschung* **19:** 105–111.

Selin, J.-F., V. Sundman, and M. Räihä. 1975. Utilization and polymerization of lignosulfonates by wood-rotting fungi. *Arch. Microbiol.* **103:** 63–70.

Shepherd, C. J., and J. R. Villanueva. 1959. The oxidation of certain aromatic compounds by the conidia of *Aspergillus nidulans. J. Gen. Microbiol.* **30:** vii.

Shimada, M. 1979. Stereobiochemical approach to lignin biodegradation: possible significance of nonstereospecific oxidation catalyzed by laccase for lignin decomposed by white-rot fungi. In *Lignin Biodegradation: Microbiology, Chemistry, and Applications.* T. K. Kirk and T. Higuchi, eds. CRC Press, West Palm Beach, Fla., 1980.

Shimada, M., T. Fukuzuka, and T. Higuchi. 1971. Ester linkages of *p*-coumaric acid in bamboo and grass lignins. *Tappi* **54:** 72–78.

Sitton, O. C., G. L. Foutch, W. L. Book, and J. L. Gaddy. 1979. Ethanol from agricultural residues. *Proc. Biochem.* **14:** 7–10.

Sjöström, E., P. Haglund, and J. Jansson. 1966. Quantitative determination of carbohydrates in cullulosic materials by gas liquid chromatography. *Sven. Papperstidn.* **69:** 381-385.

Sørensen, H. 1962. Decomposition of lignins by soil bacteria and complex formation between autooxidized lignin and organic nitrogen compounds. *J. Gen. Microbiol.* **27:** 21–34.

Sparnins, V. L., and P. J. Chapman. 1976. Catabolism of L-tyrosine by the homoprotocatechuate pathway in Gram-positive bacteria. *J. Bacteriol.* **127:** 363–366.

Sparnins, V. L., P. J. Chapman, and S. Dagley. 1974. The bacterial degradation of 4-hydroxyphenylacetic acid and homoprotocatechuic acid. *J. Bacteriol.* **120:** 159–167.

Sparnins, V. L., and S. Dagley. 1975. Alternative routes of aromatic catabolism in *Pseudomonas acidovorans:* gallic acid as a substrate and inhibitor of dioxygenases. *J. Bacteriol.* **124:** 1374–1381.

Stanier, R. Y., and L. N. Ornston. 1973. The β-ketoadipate pathway. *Adv. Microbiol. Physiol.* **9:** 89–151.

Stopher, D. A. 1960. The bacterial degradation of phenols and cresols. Ph.d. Thesis, University of Leeds, Leeds, England.

Stutzenberger, F. J. 1971. Cellulase production by *Thermomonospora curvata* isolated from municipal solid waste compost. *Appl. Microbiol.* **22:** 147–152.

Stutzenberger, F. 1979. Degradation of cellulosic substances by *Thermomonospora curvata*. *Biotechnol. Bioeng.* **21:** 909–913.

Stutzenberger, F. J., A. J. Kaufman, and R. D. Lossin. 1970. Cellulolytic activity in municipal solid waste composting. *Can. J. Microbiol.* **16:** 553–560.

Subba Rao, P. V., B. Friting, J. R. Vose, and G. H. N. Towers. 1971. An aromatic 3,4-oxygenase from *Tilletiopsis washingtonensis*—oxidation of 3,4-dihydroxyphenylacetic acid to β-carboxymethylmuconolactone. *Phytochemistry* 10: 51–56.

Subba Rao, P. V., K. Moore, and G. H. N. Towers. 1967. The conversion of tryptophan to 2,3-dihydroxybenzoic acid and catechol by *Aspergillus niger*. *Biochem. Biophys. Res. Commun.* **28:** 1008–1012.

Sugumaran, M., and C. S. Vaidyanathan. 1978. Metabolism of aromatic compounds. *J. Ind. Inst. Sci.* **60:** 57–123.

Sugumaran, M., and C. S. Vaidyanathan. 1979. Microbial hydroxylation of phenylacetic acid by *Aspergillus niger*. *FEMS Microbiol. Lett.* **5:** 427–430.

Sundman, V. 1962. Microbial decomposition of lignans. I. Identification of isovanillic acid as a breakdown product in bacterial degradation of α-conidendrin. *Fin. Kem. Medd.* **71:** 26–35.

Sundman, V., and L. Näse. 1971. A simple plate test for the direct visualization of biological lignin degradation. *Pap. Puu* **2:** 67–71.

Suter, M., R. Hutter, and T. Leisinger. 1978. Mutants of *Streptomyces glaucescens* affected in the production of extracellular enzymes. In *Genetics of the Actinomycetales*. E. Freerksen, et al., eds. Gustav Fischer Verlag, New York, pp. 61–64.

Sutherland, J. B., R. A. Blanchette, D. L. Crawford, and A. L. Pometto, III. 1979. Breakdown of Douglas-fir phloem by a lignocellulose-degrading *Streptomyces*. *Curr. Microbiol.* **2:** 123–126.

Tabak, H. H., C. W. Chambers, and P. W. Kalber. 1959. Bacterial utilization of lignans. *J. Bacteriol.* **78:** 469–476.

Tadasa, K. 1977. Degradation of eugenol by a microorganism. *Agric. Biol. Chem.* **41:** 925–929.

Terashima, N., I. Mori, and T. Kanda. 1975. Biosynthesis of *p*-hydroxybenzoic acid in poplar lignin. *Phytochemistry* **14:** 1991–1992.

Terui, G, T. Enatsu, and S. Tobata. 1961. On the dissimilative metabolism of anthranilate by *Aspergillus niger*. *J. Ferment. Technol.* **39:** 724–731.

Terui, G., T. Enatsu, and H. Tokaku. 1953. Degradation of salicylic acid by *Aspergillus niger*. *J. Ferment. Technol.* **31:** 65–71.

Thiverd, S., and P. Lebreton. 1969. Champignons lignivores. Contribution a' l'ètude de la dégradation de la lignine. *Rev. ATIP.* **23:** 313–324.

Tiedje, J. M., J. M. Duxbury, and M. Alexander. 1969. 2,4-D metabolism: pathway of degradation of chlorocatechols by *Arthrobacter* sp. *J. Agric. Food Chem.* **17:** 1021–1026.

Toms, A., and J. M. Wood. 1970a. The degradation of *trans*-ferulic acid by *Pseudomonas acidovorans*. Biochemistry 9: 337–343.

Toms, A., and J. M. Wood. 1970b. Early intermediates in the degradation of α-conidendrin by a *Pseudomonas multivorans*. Biochemistry **9:** 733–740.

Trojanowski, J., K. Haider, and V. Sundman. 1977. Decomposition of ^{14}C-labelled lignin and phenols by a Nocardia sp. Arch. Microbiol. **114:** 149–153.

Trojanowski, J., and A. Leonowicz. 1969. The biodegradation of lignin by fungi. Microbios **3:** 247–251.

Trojanowski, J., M. Wojtas-Wasilewska, and B. Sunosza Walska. 1970. The decomposition of veratrylglycerol-β-coniferyl ether. Arch. Microbiol. Pol. Ser. B. Microbiol. Appl. **2:** 13–22.

Updegraff, D. M. 1971. Utilization of cellulose from waste paper by Myrothecium verrucaria. Biotechnol. Bioeng. **13:** 77–79.

Vance, C. P., T. K. Kirk, and R. T. Sherwood. 1980. Lignification as a mechanism of disease resistance. Ann. Rev. Phytopath. **18:** 259–288.

Vidal, G. 1969. La de'gradation assimilative d'acides aromatiques par Trichoderma lignorum (Tode) Harz. Ann. L'Inst. Pasteur **117:** 47–57.

Virtanen, A. I., and J. Hukki. 1946. Thermophilic fermentation of wood. Suom. Kemistil. **19:** 4–13.

Wani, S. P., and P. A. Shinde. 1977. Studies on biological decomposition of wheat straw. I. Screening of wheat straw decomposing micro-organisms in vitro. Plant and Soil **47:** 13–16.

Watkins, S. H. 1970. Bacterial degradation of lignosulfonates and related model compounds. J. Water Poll. Contr. Fed. **42:** R47–R56.

Weinstein, D. A., and M. H. Gold. 1979. Synthesis of guaiacylglycol and glycerol-β-O-(β-methylumbelliferyl) ethers: lignin model substrates for the possible fluorometric assay of β-etherases. Holzforschung **33:** 134–135.

Weinstein, D. A., K. Krisnangkura, M. B. Mayfield, and M. H. Gold. 1980. Metabolism of radiolabeled β-guaiacyl ether linked lignin dimeric compounds by Phanerochaete chrysosporium. Appl. Environ. Microbiol. **39:** 535–54.

Westermark, U., and K-E. Eriksson. 1974a. Carbohydrate-dependent enzymic quinone reduction during lignin degradation. Acta Chem. Scand. B **28:** 204–208.

Westermark, U., and K-E. Eriksson. 1974b. Cellobiose: quinone oxidoreductase, a new wood-degrading enzyme from white-rot fungi. Acta Chem. Scand. B **28:** 209–214.

Wheelis, M. L., N. J. Palleroni, and R. Y. Stanier. 1967. The metabolism of aromatic acids by Pseudomonas testosteroni and P. acidovorans. Arch. Mikrobiol. **59:** 302–314.

Yajima, Y., A. Enoki, M. B. Mayfield, and M. H. Gold. 1979. Vanillate hydroxylase from the white rot basidiomycete Phanerochaete chrysosporium. Arch. Microbiol. **123:** 319–321.

Yamamoto, S., M. Katagir, H. Maeno, and O. Hayaishi. 1965. Salicylate hydroxylase, a monooxygenase requiring flavin adenine dinucleotide. Purification and general properties. J. Biol. Chem. **240:** 3408–3413.

Yeck, R. 1979. Methane for biomass. Environment **21:** 28–29.

Yuasa, K., K. Ishizuka, S. Kabruraki, and T. Sakassi. 1975. Metabolism of phenylalanine in Aspergillus sojae. Agric. Biol. Chem. **39:** 2199–2206.

Zadražil, F. 1977. The conversation of straw into feed by basidiomycetes. Eur. J. App. Microbiol. **4:** 273–281.

vanZyl, J. D. 1978. Notes on the spectrophotometric determination of lignin in wood samples. Wood Sci. Technol. **12:** 251–259.

ADDENDUM

The following significant papers appeared too late for discussion in the text.

Akin, D. E. 1980. Attack on lignified grass cell walls by a facultatively anaerobic bacterium. Appl. Environm. Microbiol. **40:** 809–820.

Connors, W. J., S. Sarkanen, and J. L. McCarthy. 1980. Gel chromatography of association complexes of lignin. Holzforschung **34:** 80–85.

Deschamps, A. M., and J. M. Lebeault. 1980. A survey of xylan degradation by wood-decaying bacterial isolates. Eur. J. For. Path. **10:** 316–319.

Eggeling, L., and H. Sahm. 1980. Degradation of coniferyl alcohol and other lignin-related aromatic compounds by *Nocardia* sp. DSM 1069. Arch. Microbiol. **126:** 141–148.

Federle, T. W., and J. R. Vestal. 1980. Lignocellulose mineralization by arctic lake sediments in response to nutrient manipulation. Appl. Environm. Microbiol. **40:** 32–39.

Haars, A., and H. Hüttermann. 1980. Function of laccase in the white-rot fungus *Fomes annosus*. Arch. Microbiol. **125:** 233–237.

Haider, K., and E. J. Kladivko. 1980. Transformation of [^{14}C] and [^{35}S] labeled lignosulfonates during soil incubation. Soil Biol. Biochem. **12:** 275–279.

Higuchi, T. 1980. Biochemistry of lignification. Wood Res. **66:** 1–16.

Katayama, T., F. Nakatsubo, and T. Higuchi. 1980. Initial reactions in the fungal degradation of guaiacylglycerol-β-coniferyl ether, a lignin substructure model. Arch. Microbiol. **126:** 127–132.

Kirk, T. K. 1980–81. Degradation of lignin. In, *Biochemistry of Microbial Degradation*, D. T. Gibson (ed.), Marcel Dekker, NY.

Maccubbin, A. E., and R. E. Hodson. 1980. Mineralization of detrital ligno-celluloses by salt marsh sediment microflora. Appl. Environm. Microbiol. **40:** 735–740.

Martin, S. B., and J. L. Dale. 1980. Biodegradation of turf thatch with wood-decay fungi. Phytopath. **70:** 297–301.

Meyers, P. A., M. J. Leenheer, and K. M. Erstfeld. 1980. Changes in spruce composition following burial in lake sediments for 10,000 yr. Nature **287:** 534–536.

Norris, D. M. 1980. Degradation of ^{14}C-labeled lignins and ^{14}C-labeled aromatic acids by *Fusarium solani*. Appl. Environm. Microbiol. **40:** 376–380.

Rast, H. G., G. Engelhardt, W. Ziegler, and P. R. Wallnöfer. 1980a. Veratrylglycerol-β-phenylethers, model compounds for soil-bound pesticide residues. Naturwissenschaften **67:** 404.

Rast, H. G., G. Engelhardt, W. Ziegler, and P. R. Wallnöfer. 1980b. Bacterial degradation of model compounds for lignin and chlorophenol derived lignin bound residues. FEMS Microbiol. Lett. **8:** 259–263.

Reid, I. D., and K. A. Seifert. 1980. Lignin degradation by *Phanerochaete chrysosporium* in hyperbaric oxygen. Can. J. Microbiol. **26:** 1168–1171.

Rosenberg, S. L. 1980. Patterns of diffusibility of lignin and carbohydrate-degrading systems in wood-rotting fungi. Mycologia **72:** 798–812.

Schmidt, O., and J. Bauch. 1980. Lignin in woody tissues after chemical pretreatment and bacterial attack. Wood Sci. Technol. **14:** 229–239.

INDEX

141